50 Ways to F

*'Forget carbon footprints or ozone layers. The sheer
level of sarcasm contained in this book could destroy the planet
singlehandedly...'*
ZOE BALL & NORMAN COOK

*'A fresh approach to serious issues, this is one book about
environmental issues that people should read.'*
DAVID DE ROTHSCHILD, ADVENTURER AND ENVIRONMENTALIST

*'I wholeheartedly support this book. It's a clever way to get people
thinking about our future.'*
HARVEY GOLDSMITH CBE

'A book launch that I actually want to do the cooking for.'
GEORGIO LOCATELLI, CHEF

*'What a great, irreverent approach to this hugely challenging
issue.'*
ALEXANDER MCQUEEN CBE

*'What a refreshing read. A really amusing book with green
credentials that doesn't preach. Hallelujah!'*
PETE TONG, DJ

50 Ways to F**K the Planet

50 Ways to F**K the Planet

Mark Townsend & David Glick

Collins

Contents

Introduction..........................9

Tested on Animals 13

#1 To bee or not to bee **#2** A hard halibut to break **#3** Space invaders **#4** Chemical reaction **#5** Tusk, tusk **#6** The krilling fields **#7** Seed the world **#8** Blow me **#9** Erode to hell **#10** Seal you later **#11** A whale tragedy **#12** Water shame **#13** One helluva fungi **#14** Going ape

The Ends of the Earth 101

#15 Spruced up **#16** Con with the wind **#17** Bottom trawling **#18** Palm feeder **#19** Eau naturel **#20** Not soya good **#21** Sea of change **#22** Arrested development **#23** Green light **#24** Radiating fury **#25** Climate of fear **#26** Germ warfare **#27** Not so slick

Politically Incorrect 183

#28 Grin and beer it **#29** When Porsche comes to shove **#30** Greenwash **#31** The final frontier **#32** Appetite for destruction **#33** Flying low **#34** Nuclear wasters **#35** Going bananas **#36** Great wail of China **#37** The sex factor **#38** Pulp friction **#39** The butt stops here **#40** Warm front **#41** Emission impossible **#42** Oh my green god **#43** Cold comfort **#44** Food fright **#45** Green gas **#46** Material world **#47** Eco worriers **#48** Brothers in arms **#49** Rock squalid **#50** Blaze of glory

Resources 323
Acknowledgements 327
Index 329

The statements, comments or opinions expressed by the authors of this book are entirely their own. Every effort has been made to ensure the accuracy of information but it cannot be guaranteed. Neither the authors nor the publishers can be held responsible for the actions of any individuals, or groups, believed to be misusing the content of this book.

Introduction

We have all dreamed of living for ever. Possibly even the planet, at some point, imagined itself to be invincible. Then along came humans with their revolutionary industrial activity and started upsetting the natural order. About now Mother Nature must be wondering whether she will even reach the menopause. Reality bites, sweetheart.

So how should the average human respond to this impending doom? Three choices present themselves. First up is the path of true virtue. Your every waking decision must be factored to minimize your footprint on the planet. Don't fly. Don't flush. Pass judgment on everyone else whilst you weep over a plate of sustainable steamed spinach at the goddamn wastefulness of it all.

Then there's the middle-way. You recycle the odd beer can and wipe your behind with green loo roll. You cycle when sunny. Your conscience is salved. It's a nice, comfy way but one that is taking us anywhere but a nice place.

The final option is all about the future. In short, that there is no future. Only today. Hell, optimism went out with square wheels. Treat every day as if it is your last and one day it will be. So, put your foot down like never before, it's time to enjoy the planet. Why deny yourself its fruits? These days, self sacrifice is only for those intellectually bankrupt enough to believe they can actually make a difference. It's far too late. Earth is in the terminal cancer ward with tubes rammed up its nose. It's dying for a cigarette and so are you. Go on, light up and enjoy one last gasp together. Who says the collapse of planet Earth need be all doom and gloom? Take a look at the major corporations, the politicians, the neighbours across the way with their big cars and whirlpool jacuzzis. They're all having a laugh. They appreciate the virtue of living for the moment. Ignore the do-gooders. History will articulate their actions as no more than the final Band-aid to be slapped on the Thames Barrier as it sinks beneath the rising tide.

This book is for all those who are courageous enough to cease pretending that they are doing something worthy. It's a fifty point manifesto that's honest enough to encourage what no-one dares admit. Ostensibly, it tells you how to f**k the planet, royally, with a great, steaming rocket shoved up its overblown behind. It tells you how to murder polar bears, mangle seabeds, eradicate honeybees, torch large forests, trigger a nuclear apocalypse, spread killer germs and become morbidly obese. In addition, instructions abound for how to create the most environmentally challenged eco-fashion label, manufacture an excessively extravagant rock band, throw the party to end all parties and, of course, how to die (because immortality doesn't exist, remember?) in a suitably wasteful manner.

The guidance in this book is strictly reserved for those who are deadly serious about ruining the planet in the shortest time possible. Some suggestions require minimal effort; some you might, quite laudably, already be engaged in. Some demand like-minded participants, others require individuals with the rarefied wealth and political access only a few can boast. But do not fret; you'll be surprised at the support you'll be able to count on. There's something here for everyone.

Even if you pursue only a modest selection of the suggestions that follow, take heart from the knowledge that you have contributed to Mother Earth's mid-life crisis. In fact, you will have played a part in the most seismic chapter of her existence. Your dreams of immortality might not be realised, but your actions will change the course of history. Enjoy the party my friend, you did in fact make your mark on the Earth.

Tested on Animals
Simple ways to see off living species

To bee or not to bee

Buzz off

AGENDA

* Wipe out honeybee population
* Enjoy your picnic in peace
* Destroy countryside and crops
* Save your £1 coin for something better than a wonky trolley

It may be small but it's certainly not lacking in fertility. The honeybee is a rampant member of the insect world, visiting flower after flower in a frenzy of pollination. Humans rely upon its promiscuity for flora, fauna and food. In your efforts to totally f**k the planet, there's an easy way to eliminate this bumbling competition. Very soon the bee will not be.

What a sting
Enthralled with its immutable sense of progress, humanity seems

to have forgotten that sometimes it's the little things that matter most. Don't make the same mistake. While you must be prepared to battle against Greenpeace and outperform Hugh Hefner, make sure you don't forget that little Don Juan, the bee. At the start of the twenty-first century, civilization finds itself dependent on this single insect. One in every three tablespoons of food derives directly from the pollinating prowess of the humble honeybee, with supermarkets gleefully cashing £50 billion worth of produce a year. So long as it's flogged by the kilo, chances are that the honeybee's enviable powers of fertilization have played a part. Shop shelves would look very different in a world devoid of the services of *Apis mellifera*. Hundreds of vital crops and cereals would wither. Fruit and veg staples would fade away. There'll be no more worrying about getting your five a day then! At last, gherkin-free burgers!

Entomologists (insect nerds) warn that society has become way too reliant on the honeybee. Behind the sophisticated production lines of the world's great supermarkets, the truth is that the security of the food supply lies squarely on the honeybee's busy shoulders. Surely then, all you need do is shoulder the bees out of existence and hey presto, you've delivered a deadly sting to humanity. But surely these valuable creatures are under constant MI5 protection? Don't be ridiculous!

Oh Mighty One

As yet you can only dream about Colony Collapse Disorder, the mysterious ailment that has performed such a sterling job vanquishing America's bee population. It is a strange and abrupt disease that persuades millions of bees to abandon their queen and fly off to certain suicide. Your real money-shot enlists the services of a parasite no larger than a full stop. The size of the varroa mite belies a voracious appetite. Once its jaws are clamped to a bee's stomach, it gorges upon the blood until the host's immune system can take no more. Predictions suggest that the varroa mite is able to cause a complete species 'die-out' in as little as a decade. Ostensibly, it is the Aids equivalent for bees. Honeybees are drained in hours. Hives collapse in days. With this little buddy, your job is done in a matter of weeks. Thankfully there's no bee equivalent to the condom. No Red Cross setting up clinics in the meadows. The way ahead is clear. Facilitate varroa's global spread and you've found the fastest route to being bee-free. In theory, nothing should stop every colony succumbing to these marauding blighters. In time, earth's long-term food supply will be jeopardized, plunging the planet into civic strife and conflict.

Fight or flight
At the start of 2008, the softly-spoken types of the British Beekeepers Association decided they could take no more. Barely able to hide their hysteria, its leaders warned ministers that Britain risked 'calamitous' economic and environmental hardship if the honeybee disappeared. They were not alone in their squawking. Supermarket executives agitated privately over the future fate of this bumbling insect. Apparently, safeguarding a tiny creature vital for global cereal and fruit production falls far beyond their wiles. With one of the core underpinning elements of their business at risk, they wait nervously for the first complaints to trickle in – inadequate pollination produces the misshapen, shrivelled food that so horrifies their customers.

When varroa began ravaging Britain's hives during the Nineties, the pesticide pyrethoid was promptly administered to halt the destruction. Its use came with a strict health warning over effectiveness: ministers were told the measures would triumph for a finite period only. As it transpired, only a handful of years. After that varroa mites would become immune to man-made chemicals. And so, funding was granted to develop a biological defence that would safeguard food supplies in the future.

Dr Brenda Ball, the world's foremost expert on varroa, led a team of scientists at the Rothamsted Research Institution in Hertfordshire. There was a troubling period when it seemed that, finally, a cure for varroa might be on the horizon. Significant progress was underway when, in the spring of 2006, the government withdrew financial support. Ball's team became redundant. Her pioneering work to protect nature's pollinator remains incomplete to this day. Within months of funding being terminated, the minister for sustainable farming and food hailed an 'environmentally-friendly' initiative to encourage more British-produced fruit and vegetables. No reference was made to the fact that without the honeybee this would prove largely impossible. His omission provides a salutary, but inspirational lesson to those bent on environmental Armageddon: you can often do a lot worse than put your faith in the elected few.

The government has handed out yet another 'proceed to go' card on your journey towards bee obliteration. The bee inspection service, conceived to monitor early signs of infection in hives, suddenly found its funding halved. Research on protecting the honeybee currently stands at around £200,000, a fiftieth of their pollinating value to the economy. Matters came to a head during a fraught meeting in November 2007 between beekeepers and government officials, when the farming minister Lord Rooker confessed that he too knew the bleeding obvious. 'If we do not do anything, the chances are in ten years' time we will not have any honeybees,' he said. Despite this, the British Beekeepers Association claims that funding continues to be denied. It seems bees are a victim of classic British stoicism. Admitting there is a problem remains a far cry from actually doing anything about it.

Every other international attempt to quash varroa has yet to yield an answer. Every new pesticide leads only to a new resistance. The parasite is always one step ahead. And so, its spread continues apace. In London, the first round of colony inspections during 2008 found all of the bees were dead. Few are the places left untouched by its blood-thirsty proboscis. China has submitted. The Americas have been penetrated. Australia is exhausted. Europe has its knickers round its ankles. Recently the invasion of southern Africa began. Hawaii is rapidly becoming unique in offering concrete assurances it is a 'varroa-free' locale. We'll see.

A sticky situation

Seemingly limitless in its vision of global conquest, there appears to be little requirement to encourage the worldwide operations of the varroa mite. At the moment, it is simply a case of kicking back and watching its worldwide domination unfold. Soon, experts predict, the entire planet will be contaminated by an epidemic immune to the chemicals concocted to kill it. A virulent new strain may explain why hundreds of millions of honeybees vanished in almost half of America's states in weeks, threatening £8 billion of crops. Perhaps Colony Collapse Disorder isn't such a pipe-dream after all ...

Wild honeybees, the quintessence of British rurality and heralded by everyone from William Shakespeare to Jill Archer, are on the way out. Those little buggers you see bouncing from flower to flower are invariably imported from Europe or Australia or from colonies reared by man. While you must put up with the fact that, temporarily, sufficient quality crops can still be grown in Britain, both you and the ever-growing British varroa empire can thank the government for opening the door to foreign infestations.

The demise of the honeybee has coincided with a 30 per cent increase in fertilizer use. It is no coincidence. Supermarkets, after all, must somehow compensate for a loss in natural fertility. This can only accelerate the extinction of the honeybee; the wax in beehives doubling as a peculiarly potent sink for airborne toxins. The chemicals, as well as poisoning the bees, also kill off the flowers that provide the honeybees' food. Beyond the farmers' fields the meadows are starting to look depressingly sterile. Research confirms that wildflowers like the clover and dandelion are dying in tandem with bees, their mutual dependency dragging one another to the grave. Valentines will be a cheap affair this year.

TO BEE OR NOT TO BEE   19

Oh Be-hive!

As long as governments pontificate on taking varroa seriously, there is only one winner. According to the Cardiff-based International Bee Research Unit, the one hope involves the genetic breeding of a new generation of honeybees, with jaws strong enough to yank the mites off their bodies. 'With what funding?' you may snigger. Evolution is all out of time.

There was a time when the distant hum of the honeybee was as sure a signal of summer's onset as traffic jams on the M5. These days you can enjoy your picnics and beer gardens free from their monotonous droning. While you sup your Guinness, varroa does the dirty work. Once, their sting was a childhood rite of passage, but now you can save yourself the trip to the pharmacy.

WHAT'S THE DAMAGE?

* Mysterious disease suddenly eradicates varroa parasite. Honeybee saved at the final hour. **Unlikely.**
* Miracle cure for varroa discovered by maverick oddball scientist. **Slim possibility.**
* Pioneering breakthrough discovers natural alternative to the wild honeybee's pollinating prowess. **Yeah, right.**
* The value of the honeybee is belatedly recognized by the government. Generous funding to protect the species is immediate. **Unforeseeable.**
* Varroa runs riot. Hawaii finally succumbs in late 2012. Three years later, remaining bee farms inside high-security sealed factories are infiltrated. **Anticipated.**

Likelihood that wild honeybee is extinct by 2020: 72%

A hard halibut to break

Fins ain't what they used to be

AGENDA

* Upset the natural order of the seas
* Stake out the salmon
* Free the finned-ones
* Interbreed and weaken the species

When it comes to wanton ecocide, it's sometimes good just to lay a marker, to show the world who's boss. And there are few better species with which to demonstrate your superiority in all matters ecocidal than one so finely developed as the wild salmon. Until now, these marvels of evolution have always allowed instinct to guide them thousands of miles across open waters in a current-defying voyage to their spawning grounds. But survival of the fittest? Pah! At last, man has perfected the means to subvert the natural order.

Breeding frenzy

The first step was to fish wild Atlantic salmon practically to the point of exhaustion; the second to begin farming replacement fish. And therein lay the evil genius of the plan, the 'extinction vortex'. The new man-reared specimens were inferior in all ways but one – they had what it took to destroy their wild friends. Covertly released with the excuse of having 'escaped' from farms, their mission was twofold: to breed with their genetically superior wild cousins, and then to infect them with disease. Talk about eliminating the competition. The torpedo-like physique of wild salmon – in some ways the SAS of the aquamarine world – became weakened by intermingling with the flabby farmed types, the genetic equivalent of a couch potato with fins, and the salmon's instinctive ability to survive in the wild was shot. No longer could it make its trans-Atlantic migration back home. It was fin-ished.

Born to be wild

Under cover of darkness, the men bobbed towards the vast sea cages. As their dinghy pulled alongside the expanse of steel mesh, the balaclava-wearing figures on board grimaced at the writhing coil of bodies. At a silent signal, they began to hack at the cages with steel-cutters. That September night, 15,000 halibut were liberated from their underwater prison at Kames Marine Fish Farm, Oban, off the west coast of Scotland. It was a textbook 'release'. Police were left floundering with but a single clue: the letters ALF daubed on a nearby wall. No one was ever caught.

In many respects, the Animal Liberation Front represent everything you probably can't be bothered with. Their abiding philosophy is, after all, to draw attention to and to condemn 'speciesism', an assumption of human superiority leading to the exploitation of animals. Yet these are emancipated times; tribal loyalties and prejudices have no place in the quest to deliver environmental catastrophe. To satisfy such lofty ambitions, you must adopt the same methods, if under quite a different agenda.

The guiding principles of the ALF are to 'liberate animals from places of abuse', such as fish farms, and to 'inflict economic sabotage' on those who profit from caging creatures. These two tenets fit nicely with your task to mass-release farmed salmon into the oceans of the world.

Establishing contact with militant wings of the ALF is challenging, but surmountable so long as you do not make the mistake of explaining that your real motive is to rid the seas of wild salmon, dress head to toe in khaki, and take out a subscription to the *Socialist Worker*. The actual act of liberation is no doubt a more important concern to animal activists than the genetic carnage they unleash upon the world in setting free caged animals. Despite this, it is best to err on the safe side and keep the master plan secret for as long as possible.

The ALF is a loose network of autonomous cells and in order to meet like-minded members you will need to join demonstrations against animal-research centres, trawl internet message boards or subscribe to its newsletter. But rest assured, the ALF are out there, with the organization describing its members as including 'PTA parents, church volunteers, your spouse, your neighbour or your mayor'. High-profile members are usually under police surveillance and officers typically video demonstrators at protests. A new face might attract unwanted attention. Also, do be prepared for the possibility that even ALF sympathizers may refuse to sabotage fish farms. If this is the case, don't just give up. Seek out the Animal Rights Militia (ARM) or the Justice Department, who believe that direct action is the way to go. If these two underground over-the-top movements prove too elusive, try the Lobster Liberation Front, which has already attacked fishing interests with varying success and might be persuaded to broaden its target base. Certainly, there should be enough activists around who possess the necessary zeal and wile to successfully liberate fish. After all, past attacks have proved that activists have the determination to navigate freezing waters at

 A HARD HALIBUT TO BREAK

night, the strength to cut through heavy netting and the guile to evade security.

Let's go fishing

Enticing possible recruits will rely on propagating several set arguments. Make sure you encourage rumours that farmed salmon are being obscenely crammed into cages and force-fed processed proteins by machine. Don't forget to mention the colourings and antibiotics which are prophylactically tipped into cages. Refer to caged fish as 'the battery chickens of the sea'. Animal activists will be unable to resist such bait. If more persuasion is required, refer to a past Scottish Executive salmon report which reveals that farmed salmon are recognizable from their small heads, deformed bodies, and diseased gill covers. If this fails, concoct a story about how they are electrocuted with prods if they don't finish their tea.

Once you have netted recruits to your new splinter cell, Operation Bite Back IV can get underway. Farms holding halibut and cod are all well and good, but the best objective is to disable the eighteen vast salmon farms dotted along the Irish and Scottish coasts. Further afield, the ALF's international network can concentrate on targeting the massive fish farms of Norway, home to the world's biggest salmon-farming industry. This is where the blueprint for large-scale salmon releases was written. Among a series of triumphant attacks on fish farms was the release of a hundred thousand salmon from a major facility in northern Norway, in the course of which activists slashed seventy thick nylon ropes. Comfortingly, some Norwegian streams are already populated entirely by descendants of farmed fish, fragile creatures whose presence is sustained only by the continual release of domesticated specimens. If you are forced to operate domestically, bear in mind that farmed salmon released from a Scottish fish farm can make the journey for you, swimming the distance and unleashing widespread genetic havoc in Norway itself. Most second-generation hybrids die in the first few weeks as a result of genetic incompatibilities, but researchers have found Scottish escapees flapping their fat forms all the way to Scandinavia.

See ya, Salmon

If you are to fully achieve your objectives, there are a couple of things to bear in mind. First, ensure that the salmon you release are sexually mature and capable of breeding with other fish. The last thing you need is millions of frustrated farmed smelts flabbing out and admiring their trim wild types from afar, but unable to do owt about it. Secondly, time co-ordinated attacks on the sea farms to fall between September and November when the local wild salmon are spawning. Farmed fish might carry a few extra pounds, but this at least can equip them to bully young pure-bred salmon out of the best spawning spots of rivers. Job done, they then head out to sea, never to return.

With a good wind behind you, Operation Bite Back IV will be recorded in history as achieving the biggest ever release of man's salmon progeny. The potential is staggering. Already, between 2005 and 2007, 70 million smelts were squeezed into sea cages around Britain. Up to two million farmed salmon are estimated to have been released so far this century due to accidents and the battery of rough seas. And up to 90 per cent of salmon returning to rivers in Ireland, Scotland, the Faroe Islands, Norway and Canada are already believed to be fugitives of farmed origin.

WHAT'S THE DAMAGE?

* Security measures at fish farms stepped up on police advice after series of attacks cause a few modest releases. **Certainty.**
* New virulent disease destroys farmed-salmon populations across the world. Bad news turns to good when mystery parasite then begins assailing wild populations. **Likely.**
* Farming of salmon declines in favour of farmed cod. Again, grim tidings lead to joy when move leads to resumption of wild-salmon fishing. **Plausible.**
* Animal activists fail to significantly disrupt fish farms because of new policing powers and tougher legislation to deter such activities. **Probable.**
* Despite lack of successful militant action, continued number of escapees from fish farms blamed for denuding wild-salmon numbers. Population falls to 'critical' levels within ten years. **Strong possibility.**

Likelihood of wild Atlantic salmon being extinct by 2015:
61%

Space invaders 3

The root of all problems

AGENDA

* Savage gardens
* Put down monstrous roots
* Rupture the infrastructure

The aliens landed some time ago. For a while they kept themselves to themselves and even seemed relatively well behaved. But, in truth, they were biding their time, waiting for the moment when world domination could begin. Naturally, being aliens, they would first have to morph into something terrible. And so they became a Triffid-type monstrosity, a rapacious superweed replete with superpowers. They became indestructible.

Knotted up
The Japanese knotweed, brought back to Victorian Britain from the Orient as an ornamental delight, is probably your favourite plant. A splendid-looking *pièce de résistance* with the armoury, faculties

and, most of all, ambition to subvert Europe's existing ecosystems. Knotweed is unstoppable. Labelled 'unbelievably strong' by the government's admiring Environment Agency, it can burst through concrete pavements and tarmac and topple brick walls. Floorboards have been ruptured. Roads have been split. And now, the knotweed has set its sights on the rape of Europa. More dangerous, according to Britain's leading scientists, than anything they have created with genetically modified organisms, knotweed is the second gravest threat to Europe's plants (beaten only, and marginally, by reinforced concrete). She – the invaders hail from a single female ancestor – is a fabulous, wily specimen, capable of reproducing effortlessly on her own. And she is in a hurry, with each clone capable of growing a metre a week. Horticulturists, almost hysterical with shock, claim to have actually seen her grow.

Out in the wild, knotweed has no natural enemies. Only man stands in her way and, quite frankly, he just doesn't cut it. Despite desperate and repeated efforts, nothing has been found to tame the knotweed. Trips to Japan to find a solution have yielded little. Hopes that voracious aphids and fungal rust may work crumbled long ago. Even supposedly impermeable mats laid on land have been, literally, punctured. The government is panicking. This problem plant costs nothing to spread but millions to defend against. Officials have spent more than £1.6 billion, 170 times the amount allocated to their biodiversity plan, but have got nowhere near the root of this knotty problem. Specialists can charge £40,000 to clear 5 square metres of the weed. Such is the concern that the government has recently started treating it on a par with nuclear waste. The removal of a solitary plant resembles a military operation. The Environment Agency, petrified of this ingenious nemesis, has produced a 37-page knotweed manual, which recommends digging away an area 7 metres around each plant and 3 metres deep: almost 600 cubic metres. The specimen should be removed and incarcerated 5 metres deep at a licensed landfill site. This is the only way to kill her for sure, but it has become so expensive and time-consuming that no one can be bothered. Call it natural selection, call it botanical genocide, call it what you will: the day of the Triffids is getting closer. You will hasten that day, helping this nefarious weed to overrun Europe, and sending indigenous species fleeing for cover.

Rooted in the land

Good day, Earthlings. Another Monday morning in 2017, the start of another working week under the occupation. The traffic bulletin offers a round-up of the usual pandemonium. Gridlock again on the M25 due to a weed burrowing beneath the fast lane. Near Doncaster a derailed train lies on its side after subsidence caused by a rampant plant. In the streets, commuters trudge to work in the shadow of towering stems that have pushed up through the pavement. Everywhere, the city's streets are avenues of solid, swaying greenery. In this twilight world, cars flash past with headlamps on at midday. The news brings little respite. A school in Wales has been crushed by a falling wall, pushed over by an untamed tendril. Knotweed has burst into the House of Commons, this time directly through the speaker's chair. The Queen is reportedly throwing a hissy fit because the Buckingham Palace herbaceous borders – the most heavily defended flowerbeds in the UK – have, again, been overrun.

Spreading the knotweed is child's play, but the plant's destructive tendencies ensure ultimate satisfaction. All you need do is ferry some cuttings about the continent and scatter them liberally whenever and wherever the mood takes you. Unarguably, this is one of the most straightforward means of defacing the planet. The challenge lies in blanketing an entire landmass in her shade, the creation of the first monocultural continent. This is the true meaning of going green.

Evidence indicates that Europe's entire collection of indigenous fauna and wildlife could not survive a knotweed kingdom. When the plant wrested control of a Cornish valley in 2007, choking the

SPACE INVADERS

landscape with a 7-mile bank of weed, naturalists recorded a mass exodus: dippers, grey wagtails, Daubenton bats, bluebells, and thrift all scarpered. Even the yellow flag surrendered without having time to change hue. Species have a choice; they either fight or flee. And the recent past shows that the former is futile.

Back to the roots

To get Europe knotted you will first need to locate the weed. This won't be too problematic. Already she has spread from Land's End to the northern tip of the Isle of Lewis, her striking good looks immediately noticeable; a touch of bamboo bristling with fluffy white flowers and orangey-yellow roots, quite fetching on every level. Only the Orkneys have escaped so far. Scour rubbish tips or derelict land; deserted places where you won't be disturbed. If you are, merely pretend to be a good citizen cutting down the ubiquitous weed (cutting or mowing encourages its spread, but you will conveniently forget to mention that bit). Knotweed spreads using its rhizomes – its roots – and a fragment as light as 0.08 grams – fingertip size – is all that is required to grow another plant. With her labyrinthine roots encompassing an area the size and depth of a subterranean swimming pool there is no shortage of incendiary material.

Fill several dozen bin bags with rhizomes and place in the back of a truck with blacked-out windows. Inside the truck, start shredding the roots into pea-sized pieces, a tedious process offset by the knowledge that each tiny shred is sufficient to start a fresh colony in a location of your choosing. As you leave, be sure to drive over the dig site; scraps of knotweed stuck to tyres have, in the past, facilitated cross-country transfer with triumphant results.

Sow the seed

Now for the fun bit. The list of attack sites is innumerable. Some are fairly obvious but, really, it's up to you. Go crazy. National Parks are particularly fair game as, clearly, is any site considered naturally exquisite. Thrill-seekers might want to share their cargo with the grounds of Balmoral or Buckingham Palace. Prince Charles's organic estate at Highgrove exerts a certain pull, as does the prime minister's country residence at Chequers. Catapults armed with pellets of knotweed rhizomes and weighed down with pebbles seem an obvious tactic for penetrating such hallowed grounds. Maybe consider a remote-controlled plane with remotely activated fuselage doors to release knotweed bombs. Target the gardens of folk like Alan Titchmarsh, whose penchant for televised botany more than justifies such actions. The world-famous Royal Botanic Gardens at Kew. The Royal Horticultural Society's showcase at Wisley. The list is endless. Scatter knotweed roots into the Thames to float downstream and impregnate the banks. They might even drift out to sea and contaminate some faraway land. Risk her remains on the playing fields of major sports stadia – Anfield, Aintree or central court at Wimbledon. London's 2012 Olympic site would have been another indisputable target, but a welcome infestation has already swamped 10 acres of it, amid reassuring reports that it may even delay the games. Hire a hot-air balloon and float above the countryside with several kilos of chopped knotweed for company. Elevate to several hundred metres and release 37,000 snippets of rhizomes at five-minute intervals. Why not head abroad? Enough firearms and drugs are smuggled into Britain every year for you to be certain that a few sprigs of weed root will get through. Sniffer dogs aren't trained to search out knotweed. Once on the mainland, you know what to do. Incidentally, the precise technique for dispersal is not really an issue. Either hurl into the air and let the weed herself decide where she lays her roots or place firmly into soil. Any soil will do; even the most wretched quality is sufficient for this hardy little sprig.

   A word of warning: it is illegal to propagate or even transport knotweed. Although we don't want to encourage law-breaking, sometimes the ultimate goal – that of completely f**king over the environment – must take precedence. Anyway, the sentence is soft

 SPACE INVADERS

in light of the potential rewards. On the slim chance that you get caught – and as yet no cases have come to light – you will face a maximum two-year sentence. With characteristically good behaviour, you'll be out in a few months, just in time to witness the first shoots of your labour, before being caught hang-gliding with a sack of knotweed cuttings above the Blue Peter garden.

WHAT'S THE DAMAGE?

* Invasive species with predilection for eating knotweed is introduced by government, eradicating the weed within two years. **Never.**
* Japanese knotweed replaces rose and thistle as official emblems of England and Scotland. The EU adopts it as a symbol of unity. **Unlikely.**
* Council sued for manslaughter after child disappears down hole in playground caused by knotweed. Lawyers argue that officials displayed sufficient negligence by not heeding warnings. Knotweed control becomes pivotal issue during 2012 UK elections. **Possible.**
* Knotweed arrives on Orkney in summer of 2010. Arrival is traced back to climate-change charity walker relaxing after traipsing to John o'Groats on tedious walk to raise awareness. **Credible.**
* New superhybrid of Japanese knotweed, giant hogweed and the dreaded Russian vine is discovered. Tabloids dub it 'Invasion of the Killer Knotweed II'. **Bring it on.**

Likelihood of knotweed colonizing most of Europe by 2020:
78%

Chemical reaction

Hormone treatment for all!

AGENDA

* Dole out the contraceptive pill
* Turn sealife female
* De-fertilize fish
* Trout off the menu

Maybe the world will not end with a bang after all, but with a whimper. Instead of Armageddon and its attendant boiling seas and titanic ructions, maybe we'll just finish up trapped in a unisex world, wondering where the next generation will come from. Already, Mother Nature has started the ball rolling. The feminization of wildlife is well underway amid welcome warnings that this could dismantle an evolutionary process which has taken 3.5 billion years to perfect.

You must aid the process. The experiment will start with fish and your plan is to transform all male freshwater fish into females,

CHEMICAL REACTION

a move that will prove to be a less than progressive step for the future of the fish population. Although tests on fish breeding patterns are relatively rare, consensus and common sense dictate that making any species all female will have a profound effect on reproductive patterns. The sexual emancipation of the human female has handed you the perfect weapon. Millions claim that the contraceptive pill is a blessing. Not many expected it to prompt an environmental crisis.

The bitterest pill

Just above the West Yorkshire town of Castleford, close to the banks of the slow-moving River Aire, protrudes a pipe. Passers-by spare barely a second glance for yet another sewage outfall. They should take more notice. Or at least the blokes should. They are witnessing the cusp of the new sexual revolution. Within these brackish waters something odd is happening to fishing tackle, and we're not talking rods and floats. The Aire's male fish are turning into women, with tests indicating that 100 per cent of male fish show evidence of feminization.

The nondescript pipe above Castleford is dispensing, quite literally, the waste of humanity. In West Yorkshire, like in most places, quite a few women take precautions, and so their urine contains the female hormone oestrogen. It seems the fish here have been force-fed the female contraceptive pill. Over time, the males have begun to grow female reproductive tissues and organs. Parts of the testes turn into ovary tissue or, if they are really unlucky, development of the fish's manhood is merely retarded. In lowland parts of the river, the government's Environment Agency noticed up to half of the male fish developing eggs. Tests around the world reveal that even the tiniest traces of synthetic female hormone are sufficient to corrupt wild fish populations. Some scientists even suggest that the concentrations sufficient to make fish unisex are below detection limits in place for drinking water.

Your task is to give aquatic males the world over a helping hand in their quest to become women, albeit against their wishes. For this you will require supplies of the synthetic oestrogens widely used in the Pill. The obvious choice is etinyloestradiol, one of the most common components of the contraceptive and up to a hundred times more powerful than any naturally occurring oestrogen. Its potency is enormously reassuring. Medical advice for male-to-female transsexuals dictates that the stuff offers not only a long 'half-life and high potency' but, more importantly, gives 'excellent feminizing effects'. Traces of the Pill have been found in waterways at dosages of one part per billion. If you can only double that, you will be assured of success in your aquatic sexualization experiment. Etinyloestradiol is made in industrial quantities and at face value costs less than 8 pence for a month's supply. Several large UK companies manufacture etinyloestradiol in their laboratories. Around £10,000 should buy you enough to emulate the effects of 43,000 women taking the Pill, but for one day only. More than 3.5 million women take it every day in the UK, with 100 million worldwide. Clearly, you alone cannot afford to mimic the entire population of British women but, as a start, it will do. Be warned, though: you may have to justify your excessively large etinyloestradiol order by pretending that you are an NHS supplier.

U-bender

The next step is to target the stuff where it will have the most effect. And that, as the West Yorkshire pipe amply reveals, is perhaps the easy part. Just flush it down the loo. Then, contact like-minded ecocide sympathizers, pass on details of where to obtain the chemical and soon you will command a small network of UK volunteers stationed by their lavs in the pursuit of aqua gender-bending. It might be prudent to dissolve the pills in warm body-temperature water before flushing, in order to guarantee that they safely navigate the antiquities of Britain's sewage system. Thankfully, the effects of etinyloestradiol will not be diluted.

 CHEMICAL REACTION

Conventional sewage treatment does not eradicate the hormone, and synthetic oestrogen is not broken down in the wild, a factor that grants it better weighting on the value-for-money scale. The effects will be reassuringly quick.

Scientists in Canada added oestrogen at levels found in sewage to a remote lake in Ontario. After just a year they started to observe a creeping feminization. Even the male fathead minnow (and one wonders in disbelief how this creature could ever hope to pass itself off as a lady) turned sex. Inevitably, delightfully, 'reproductive failure' followed and fathead numbers began to crash. They never recovered, according to the results of the Canadian tests.

Other inspiring reports abound. There are the male tadpoles in Sweden who morphed into females after being fed oestrogen. In one experiment, tadpole dudes who were fed heavy dosages of the hormone became 100 per cent dudess. An Environment Agency report tells of roach who, after feasting on oestrogen, experienced deformities in their sexual organs and began producing eggs rather than sperm. Results of tests on zebra fish at Cardiff University using etinyloestradiol were so pronounced that researchers expressed unease at observing 'large-scale effects at such low levels of concentration'. But the best news arrived with reports that a male hornyhead turbot had been transformed into a lady. The development must have caused groans in rivers from Leeds to Lagos. If the macho hornyhead could be tamed, scaly chaps everywhere must have thought, what hope for them? The game was up.

Chemical cocktail

Of course, there are other ways to turn man to woman, and it would be unwise, even reckless, to overlook the old 'gender benders', or 'endocrine disruptors', as the scientific community would rather these chemical lovelies be known. Synthetics found in plastics, shampoos and food packaging mimic oestrogen when ingested. Such useful material gets everywhere and is particularly effective when ingested in cocktail form over the years. Plenty of these 'endocrine disruptors' appear to be travelling north on the moist air currents that blow from Europe to the Arctic. When confronted with the frosty, Arctic air, these chemicals condense and fall. They are perhaps most revered for creating the famously pseudo-hermaphrodite polar bears with penis-like stumps, a result which to this day remains one of the most celebrated achievements in the world of chemical scalps. And clearly there is no merit in trying to better such a masterpiece.

Despite a forensic EU review of chemical legislation, a decent number (up to five hundred) of potential endocrine disruptors remain in use, and, thankfully, these hormone-disrupting chemicals are still allowed to be sold even though safer alternatives are available. The next review of legislation is not due until 2012, giving you a reasonable window of opportunity to shrink the man bear a little more. By then, who knows what else will have shrunk, changed or grown?

Let's hear it for the clam

When scientists chose an estuarine site in the West Country to examine clams, which they had naively hand-picked in the hope that they would be free of chemicals, they experienced a bit of a shock. The clams were transsexual, their Devonian testes containing both sperm and eggs. 60 per cent were like this. For you, it offers a thrilling possibility: initial evidence that oestrogen can potentially survive in seawater, battling the salty tides to turn aquamarine life female. Here, quantities of the female hormone had messed around with clams at the bottom of the food chain. Could, you dare dream, the entire sea one day be declared girls only? Embarrassing speedos would become a thing of the past.

Little surprise that those studying the feminization of fish describe the West Country clam as the aquatic equivalent of the miner's canary, the bird which chirps an alarm long before men are aware of impending doom. With the mines now mostly closed, instead there's something in the water. The sexual revolution has

CHEMICAL REACTION

proved to be intensely liberating for humanity. Now it is the turn of the misunderstood bloke fish who only ever wanted to be a woman. Once we start adding the Pill to the rivers of the world, there is no reset button. The new sexual revolution is underway.

WHAT'S THE DAMAGE?

* Government makes over-the-counter pill more widely available. **Certainty.**
* Male pill goes on general sale. Female carp duly start growing penises. **Unlikely.**
* Etinyloestradiol leak reported from factory into river system. Male fish in the area seen wearing matching bra and knickers. Scientists say this could be evidence of feminization. **Unlikely.**
* Male bull whale gives birth after growing ovaries. **Never.**
* Overhaul of the disposal of the contraceptive pill announced by government. From 2011 it has to be disposed of in special council bins and safely stored in landfill sites. **Possible.**

Likelihood of majority of male fish turning female by 2015:
23%

Tusk, tusk

Don't be a Dumbo: sign up for safari

AGENDA
* Forget the elephants
* Raise the price of ivory
* Tally ho, trophy hunters!

Considered a deity by some, treated like royalty by others, earth's largest land animal is as revered a target as you're likely to find. Environmentally, the African elephant is nauseatingly virtuous. Even its nutrient-rich faeces are laced with good ingredients for foragers, and some hapless seeds can't germinate unless they have first travelled through the elephant's vast bowels. Conservationists gloss over the creatures' destructive tendencies: they decimate crops, trample livelihoods and can gore people to death. They might look impressive, but can an elephant earn a million? Complete the Rubik's cube? Invent the iPod? No. The elephant is nothing but a jumbo Janus with teeth that do not fit in its mouth.

Tooth or dare
It is these teeth that will ensure the downfall of the African elephant. At the height of the ivory rush, elephant numbers

slumped from 1.3 million in 1979 to 625,000 a decade later. Trinkets made from their molars were highly desirable and poachers were cashing in. In 1989 some fusty chaps from the UN Convention on International Trade of Endangered Species banned the sale of ivory. The slaughter stopped pronto.

Now, though, poaching is back big time. A one-off sale from the ivory stockpile of South Africa, Namibia, Botswana, and Zimbabwe was recently sanctioned by the UN. The move has paved the way for an elephant massacre. No one can really distinguish between legal and illegal ivory. Hidden behind the smokescreen of this legally sanctioned sale, you could bring about the greatest elephant holocaust ever seen and flood the world market with illegal ivory. The price will soar. Currently, ivory fetches around £375 per kilo, up from £50 a few years back, but you must push it to a record high. The higher the price, the more likely that organized poaching syndicates will be tempted. You could try to beat the cost of gold – at the time of writing a record £500 an ounce.

Between 470,000 and 690,000 elephants are still stomping across the African savannahs. Latest intelligence suggests that 23,000 elephants a year are currently being killed for ivory. Credible certainly, but too modest a tally for your purposes. Make those boys back in the Seventies look like the cowboys they were. Their chaotic, frenzied spree massacred 70,000 creatures a year; but there's no excuse for inefficiency. With a sophisticated, organized poaching network you can hit six figures.

Safari, so good

To commence the cull you clearly need to target a country with a sizeable elephant population. Ideally, choose an economy which has collapsed beyond recognition, a country without moral leadership, dominated by a crackpot dictator with no interest in or concern for what the world thinks ... off the top of my head, Zimbabwe. But you will have to move fast. Zimbabwe's elephant population of 120,000 may not survive much longer. Already, poaching there is out of control. Latest intelligence found 939 active poaching camps in a single north-eastern state.

The first step is to get a foothold inside this turbulent country which, while starving its own people, has retained the admirable foresight to continue allowing elephant hunting. Start by trawling safari companies, who can offer a permit to enter Zimbabwe for the 'trophy' hunting of such creatures. Shooting animals for a laugh is a traditional pastime of the privileged, so try the royal gunsmith Holland and Holland, based in Mayfair and supplier of firearms to creatures as respected as the Duke of Edinburgh, the Prince of Wales and members of the new aristocracy, Madonna and Guy Ritchie. Once, H&H charged an awesome £5,000 for shooting male elephants in Botswana. Have a punt and ask about neighbouring Zimbabwe. Admittedly, it's a long shot – they claim no longer to offer such well-meaning trips. Otherwise, try Banbury-based E.J. Churchill, who once offered elephant shooting at a not unreasonable £4,700. Company director Sir Edward Dashwood is an admirable old card, once saying of pachyderm poaching, 'Go and shoot an elephant, it's like having the most expensive bottle of wine you can have at every meal, with vintage champagne and caviar.' Although he no longer organizes these worthy African breaks, he may be able to recommend like-minded sorts who still do. If he's reluctant to help, your third port of call is Zimbabwe-based Nyakasanga Hunting Safaris, who offer professional licensed hunters to target male elephants.

Bull's-eye
Once you've made it to Zim, make efforts to shoot only bulls. Men might be increasingly redundant in human society, but they're still required to do the business out here on the African plains. Spread the word to the local poaching networks that you'll pay $1,000 – valuable foreign currency – to anyone who slays an elephant. Alternatively, contact Zimbabwe's National Parks and Wildlife Management Authority, which has reportedly been offering

TUSK, TUSK

farmers the chance to buy elephants. According to the authority's director, Maurice Mutsambiwa, they have at least 55,000 too many. The going price is £1,000. What a snip.

So, hurrah! There is at least some good news coming out of modern-day Zimbabwe. Mugabe may run the risk of being under-appreciated, but you must applaud his well-crafted food shortages which impel his countrymen to kill elephants for food and his officials to deal in ivory in order to feed their families. One national park has, according to reports, been instructed to slaughter elephants to feed the villages on Independence Day. What a feast! All indications are that your poachers will be able to operate with impunity as long as Mugabe gets his return. Zimbabwe's elephants stand no chance.

Ivory coast

The cull should next be encouraged to move on to Zambia's 30,000 elephants and then Botswana's 123,000 creatures, followed by an efficient extermination of Namibia's 12,000. Next, the 17,000 of South Africa, the 24,000 of Mozambique and, finally, on to Kenya's population of 32,000. It is now early 2012 and more than 350,000 elephants in Africa have vanished. As southern Africa is cleansed, mobs up north have extinguished the Loxodonta africana in the Democratic Republic of Congo, Cameroon, Tanzania and Nigeria where, in places, the fog of war has abetted your programme.

Unsurprisingly, the massive market of China is obsessed with ivory; its elite will do anything to get their hands on the new 'white gold'. And while Japan might be the only certified country able to receive ivory stockpiles, all illegal African ivory will be easily diverted en route to this rising star of the East. Chinese traders will facilitate the movement. Dozens of companies there are currently registering in preparation to legally sell ivory. The China National Petroleum Corporation and its Ministry of Defence have both been linked to the ivory trade. Even the Communist Party of China has been involved, which just goes to show that socialism works in mysterious ways. China is currently pressing for the ivory moratorium to be lifted. Botswana, Namibia, South Africa and Zimbabwe, meanwhile, are chasing regular export quotas. Even if they fail, it really doesn't matter as long as the teeth reach China. There they can use it how they want: ivory chopsticks, figurines, spectacle frames – you name it. All that is important is maintaining the fabulously well-formed smuggling route from the poaching camps of Africa to the shipping magnates of China.

African nations should learn from the example of Zambia. A shipment of 6.5 tonnes of ivory – 6,000 elephants' worth – was intercepted in Singapore, most likely on its way to China. When questioned, Zambia's government played dumb. Its official documentation claimed that only 135 elephants had been killed in the country during the entire previous decade. Their bluff worked, but in the future countries should at least attempt to make their story plausible.

Keep an eye on other potential markets. Europe has always been partial to ivory. Again, aristocracy may have a part to play. Down the road from Holland and Holland is the barbershop George F. Trumper Ltd, patronized by the royals since Queen Victoria's reign. The 'finest traditional gentlemen's barber in London' was once fined £10,000 for dealing illegal ivory shaving brushes and the like. Soon after, the Met realized that protecting animals was actually a bit of a lame pastime in the face of real crime, and the four-strong unit was scaled back to two because accountants felt that £40,000 from a £2.5 billion budget could be better spent. And so the ivory keeps on coming. Recent audits found 27,000 ivory items on sale in 1,143 shops across Germany, UK, France and Spain. Germany hoarded more than 16,444 items, with the UK coming in at a respectable second-place with 8,325. If all goes well, one day, quite soon, such figures will seem ridiculously, naively, low.

 TUSK, TUSK

WHAT'S THE DAMAGE?

* Well-known environmental group calls for large-scale trophy hunting of elephants with proceeds spent on conservation measures. **Certain.**
* China's president appears on state television wearing what looks suspiciously like an ivory bracelet, beads and ring. He responds to international opprobrium by explaining it was a gift from his African friends. Chinese media remarks how handsome their leader looks. **Unlikely.**
* Price of ivory reaches £1,100 in 2014 after conservation groups announce that the wild African elephant population has collapsed. A rare interception of ivory for Beijing is traced back by DNA to Mugabe's presidential herd. **Improbable.**
* As poaching spirals out of control, world leaders promise international efforts to save the elephant. Britney Spears goes slightly madder and media attention shifts stateside. The Serengeti is all but forgotten. **Probable.**
* Intent on proving he is more bonkers than Spears, Mugabe announces he cannot find any of his country's elephants. Last seen riding naked into the jungle on the back of his last surviving presidential elephant. **Feasible.**

Likelihood of African elephant reaching unsustainable levels by 2020: 71%

The 6 krilling fields

Sea into the future

AGENDA

* Kill the krill
* Vacuum the Southern Ocean
* Up your seafood intake
* Celebrate the health benefits

It is the one that most definitely got away. Amidst the perishing netherworld of the vast Southern Ocean lies the holy grail of fishermen: the krill. Little is known for certain about this bug-eyed crustacean. What is beyond dispute is that it has become the largest marine resource left on the planet. Trillions of krill bob about in these mysterious waters, an estimated 7 million for every person, and their importance to the health of the planet is slowly and increasingly becoming a matter beyond debate.

THE KRILLING FIELDS

Between a Rokke and a hard place

The krill supports practically all life in the Southern Ocean, including those lording it at the scenic end of the food chain. Among them, the mightiest animal on the earth, the blue whale; the greatest seabird, the albatross; not to mention all manner of adorable seals and dolphins. Without krill, the ecosystem of Antarctica, the planet's last unspoilt continent, would collapse and the world's fourth largest ocean would become a vast and boundless empty space. Hoovering the Southern Ocean of krill would see its freezing depths colonized with nutritionally useless gelatinous blobs called salps. An ocean currently teeming with life would become a wobbling wasteland of jelly.

Time, then, for the great krill cull. Market research suggests that these crustaceans will prove a sure-fire hit in the supermarkets. The meat contains two of the most sought-after health ingredients around, omega-3-rich phospholipids and the antioxidant astaxanthin. If krill could be supplied to those eager to self-improve, millions of the world's most health-conscious consumers, ready to embrace the newest 'superfood', would be forced to play an unwitting part in accelerating the demise of their planet. And such delightful ironies should never be passed over in your laudable quest to f**k the planet.

But there is a problem; a biological conundrum that has proved nigh on impossible to crack. Krill are a smart bunch, boasting a self-defence system that explains why this 5-centimetre-long shrimp has managed to amass the greatest populational biomass of any multi-cellular species in modern evolution. Despite decades of attempts, the krill has unerringly outwitted man's attempts to exploit its bounty. Somehow, these crustaceans had the annoying evolutionary sense to develop a unique trait – once snared in a fishing net, they immediately self-destruct. The instant they are hauled on deck, potent enzymes are unleashed inside their bodies, causing their insides to rot. In a moment their colour drains from pink to grey and their bodies dissolve into inedible mulch. It is a deeply vexing state of affairs. Not only has this self-defence mechanism ensured their abundance, it has preserved Antarctica's entire ecosystem.

All that is about to change. For the first time in history, the swift eradication of the innumerable, ingenious krill can be safely predicted. Thrilling reports have begun circulating in maritime circles of a technological breakthrough that will syndicate the end of the world's last great untapped fishery. Kjell Inge Rokke, Norway's richest man, and owner of Aker fishing conglomerate, is a man who must be courted. His firm has lodged a patent for a system allowing krill to be piped in their billions on to a trawler's deck and frozen immediately. Overnight, the crustacean has become fit for consumption. History confirms that there is nothing quite like the arrival of a new cheap fish to hasten premature extinction.

Rokke's patent is gold-dust in the currency of species annihilation. The profits to be made from this largely untouched resource are stupendous. A recent appraisal by Swedish bank Enskilda estimated Rokke's krill venture at £2.2 billion a year. And that's just the profit. Offers of investment should be forwarded directly to Rokke, who will probably sniff at any piffling amount of money you may be able to offer but might be tempted by the promise that his invention will become widespread. There are various other reasons to believe he might make himself available to listen to proposals that include monopolizing krill until complete exhaustion. After all, this is the billionaire who served a jail sentence after attempting to bribe a Swedish yacht inspector for a new fishing licence.

Rokke the boat

As always, time is of the essence, and you should aim to approach Rokke's people as soon as possible. Modification work has already started on the Aker fishing vessels. Latest intelligence suggests that refurbishment of the 92-metre trawler, the *Saga Sea*, is

 THE KRILLING FIELDS

proceeding apace. Once ready, the *Saga Sea* will be able to vacuum a record 120,000 tonnes of krill in a single fishing season. Soon afterwards, a new £85 million krill vessel will be constructed with the ability to catch almost double that amount. Calculations confirm that a 25-strong fleet of Rokke's super-trawlers, each working just one season a year, would exhaust the Southern Ocean's conservative estimate of 60 million tonnes of krill in a decade. And he has the personal wealth to fund the entire lot. Although rival fishing conglomerates are eyeing up Antarctica's pristine killing grounds and its promise of a quick slaughter, flamboyant Rokke is the man. His wealth has been largely built on fish. Let's hope the industrialist will not shirk his greatest opportunity yet.

Ideally, any business model presented to Rokke should be inspired by the scale of exploitation previously seen on the Grand Banks off Newfoundland, one of the most inspiring episodes of man's early forays into species extinction. In the sixteenth century, these Canadian waters were so brimming with cod that sailors bragged of walking to land across their scaly spines. Once the world's most bountiful fishing ground, the Grand Banks were closed in the early Nineties. Fishermen had seen fit to target these waters as if cod were inexhaustible. The cod have never returned.

Examination of the Newfoundland fishing strategy also underlines the importance of attention to detail when conducting the fine art of annihilating a species. Rokke's fleet must target the periphery of Antarctica's ice sheets during the summer months. This is the time when penguins and seals forage for the crustacean to feed their young. Life may be tough out in the freezing, dark waters of the Southern Ocean. Without krill, it is impossible. A report by the Royal Commission on Environmental Pollution warns of krill's importance in sustaining seals, whales and penguins and ensures that any krill cull will be 'deeply emotive'. Which, for your purposes, only underlines the need to get moving. To offset the risk of meddling greenies, an unswervingly misleading PR campaign blaming dwindling penguin and seal populations on the vagaries of climate change should be prepared. Particular attention should be paid to seals. People get unusually animated over their fate and, should your campaign prove successful, they will start shouting at the US or China rather than Rokke's well-trained fleet. You should also make the most of the lack of hard science surrounding the krill. Any tiresome Preserve the Penguin or Save the Seal campaigns that do make the link with gradually depleting krill stocks will be countered with bogus data challenging their reliance on the crustacean. There seems scant prospect of a queue forming in front of anyone deluded enough to find themselves rattling a Keep the Krill collection tin. A shrimp the size of a pen-top can only fail to impress in a world where only the fluffy and oversized are championed by conservationists. Give them a choice of saving either the bug-eyed shrimp or the lion and, well, down with the krill.

The PR offence should not stop there. A lobbying campaign needs to be mounted championing more liberalized krill-fishing quotas in the Southern Ocean. This is unlikely to meet with objections. The treaty protecting Antarctica promotes freedom of scientific investigation. If the Japanese can restart whale slaughtering under the pretext of academia, then mopping up an extra squillion or so krill should be a formality. The details of the Convention for the Conservation of Antarctic Marine Living Resources – the coalition which governs its seas – offers no additional concern. Structurally, the convention is perfect; the usual mishmash of competing agencies obligated to make decision by consensus. One negative vote can block a crucial conservation measure. They rarely agree. The US is just one nation who can be counted on to disagree on any matter of conservational importance. A cursory inspection of the convention's fishing regulations generates further confidence. Not only is the krill fishery exempt from the few guidelines in existence but, unlike other catches, krill trawlers are not required to submit information on their intentions. Rokke is no doubt aware of such small-print.

THE KRILLING FIELDS

In addition, trawlers are not obliged to carry satellite-tracking systems, ensuring that officials are clueless as to where Rokke's fleet are hunting in an ocean four and a half times the area of Australia. And, finally, there is the trusted black art of the fishing industry, namely, fiddling the quota.

Be confident that humans will take to krill like a fish to water. In the endless quest for a longer life people crave nothing more than another miracle food. Krill fits the bill; an antioxidant-rich shellfish straight from the unpolluted waters of the deep south. Care should be taken to advertise the crustacean as 'krab' rather than 'krill stix'. It is of the utmost importance that this tiny creature is vanquished without sinking into public consciousness. The krill cannot be blamed for assuming they would survive for ever. Unfortunately for them, their days are numbered.

WHAT'S THE DAMAGE?

* Rokke rudely rejects your offer of help and adopts a sustainable-fisheries plan. **Plausible.**
* Guidelines for krill trawlers in Southern Ocean are unexpectedly tightened. **Unlikely.**
* Krill become the 'superfood' of the fit and fashionable. By 2012, harvests cannot keep pace with demand. **Conceivable.**
* Fishermen working off Antarctica are prosecuted for the first time for exceeding quota limits. Plans to over-fish the Southern Ocean rapidly scaled down. **Little chance.**
* Krill breeding rates confound all expectations, making them impossible to eradicate. **Slight prospect.**

Likelihood that krill is extinct by 2020: 70%

Seed the world

Sow the seeds of destruction

AGENDA

* Rob the Arctic bank
* Blot out a few botanists
* Stamp on the seeds of the future
* Become an eco-martyr

On the most northerly point of earth accessible by aeroplane, man has made preparations to safeguard his existence. Here, on a remote Arctic archipelago, a secret doomsday vault is hewn into the side of a frozen mountain. Buried deep within its icy chambers are the seeds on which the future of mankind's food supply depends. This top-security repository is the planet's most important seed vault. Seeds from every crop variety on the planet will be safely stored here; the last line of defence against the extinction of agricultural diversity. Climate change has created a desperate need to store varieties that can adapt to the changing weather. The plan is simple: should a crop disappear, its seed will be plucked from the vault and germinated once more.

Noble sentiments indeed, but clearly you must destroy this modern-day Noah's ark. When you extirpate the planet, you will want to know that it is for good. What you don't want to discover is that humanity has an escape clause, an insurance policy that allows the planet to replenish itself. There must be no second chance, the seeds of this contingency plan must fall on stony ground.

Vaulting ambition

The Svalbard seed bank is considered the one complete, secure botanic gene bank in the world. Others have been looted or destroyed by typhoons. Enchantingly, most of the rest are at risk from war, fire or natural disasters. This bank is deliberately sited at what is arguably the most secure place on earth. Far from the troubles of mankind, the world's seed depository sits on the isolated Norwegian island of Svalbard, just 500 miles from the North Pole. This is not a place where eco-terrorists usually hang out and its designers claim the Svalbard Global Seed Vault can survive an earthquake, a nuclear-missile attack and all kinds of global calamity. They claim it is indestructible, the 'living Fort Knox'. Its designers are simpletons. Nothing, as you will prove, is impregnable.

While planning your sabotage, you learn with a heavy heart that, within days of opening, the vault did indeed survive a 6.2-magnitude earthquake, the biggest ever recorded in Norway. Your hopes that climate change will adversely impact the vault are similarly dashed by reports that the vault is sited 130 metres above an icy fjord to protect it from any rise in sea level. Its cocoon of permafrost is so dense that a warming climate will not make a difference for centuries. Latest security updates make similarly bleak reading. There is only one way in and out, through a narrow reinforced concrete and steel door, planted halfway up a mountain.

It requires a master key to gain entry. Oh, and there are armed guards. As the vault sits on an impenetrable frozen layer of permafrost, tunnelling from below is impossible. The use of high explosives is similarly futile; the vault is shielded on all sides by at least 150 metres of frozen rock. Cutting the island's power supply is also pointless, as the permafrost ensures temperatures inside the complex never nudge above -3.5°C, helping to ensure that the seeds remain healthy for up to two millennia. Even getting up close is difficult, due to the perimeter walls of fortified brick that overlook the single approach road. There, with polar bears doubling as sentries, it seems that nature herself is keen to safeguard the vault's valuable booty. In a nutshell, destroying the Svalbard Global Seed Vault looks bloody tricky. Yet these difficulties must be overcome. The brilliance of sabotaging the Svalbard seed vault cannot be overstated. And, unlikely as it seems, it is the polar bears that hold the key to penetrating the most environmentally prized bank of all.

Raid on

Getting to the region is easy enough. Frequent flights go from Oslo to the archipelago's one major settlement, Longyearbyen, which has a population of 2,300 and is a short drive from the vault. The timing of your arrival on Svalbard is crucial. You must ensure your vacation in Longyearbyen coincides with a fresh delivery of seeds to the vault. The first consignment of 250,000 arrived in February 2008. Another two or three million unique varieties have yet to arrive, with room for 4.5 million samples in all. Regular imports of batches are expected to be flown in from around the world. From now on you must monitor the websites of various conservation groups, such as the Royal Horticultural Society and, most importantly, the Global Diversity Crop Trust, the brains behind the project, who, along with other groups, will issue press notices to keep everyone informed of progress. Once you know the next scheduled arrival, book four airline tickets (you will

SEED THE WORLD

need a team of at least this size to successfully execute this mission). With your team based at one of Longyearbyen's hotels, your cover story is that you are in Svalbard to shoot ... erm, spot some polar bears. In actual fact, you are keeping watch for the arrival of international botanists bringing in the latest genetic safeguards. When these botanists arrive, they will probably head straight to the vault. Jump into your hired 4x4 and follow discreetly. If fortune is on your side, their arrival will be timed between November and January, when there is unbroken darkness, allowing you to follow their convoy undetected.

Keep your weapons handy. Anyone straying outdoors in Svalbard is required to carry a high-powered assault rifle in case a polar bear attempts to find its dinner. As an upstanding steward of the planet, you are merely obeying the instructions contained within the Norwegian ministry for the environment's glossy Svalbard tourist brochure, which stipulates: 'Carry a weapon when travelling outside the settlements. Shoot to kill.' You do not need reminding. Thank heavens for polar bears, for without them you'd have little excuse to be armed.

As the botanists approach the narrow bridge to the vault's entrance, they will have to unveil a master key. That is your signal. Ram the complex's outer gates and open fire on the armed guards. The element of surprise should ensure a clean victory. Accelerate towards the bridge, leap from your jeep and take the botanists hostage. With assault rifles pressed to the temples of these well-meaning plant scientists, force them inside the vault, leaving two of your team to guard the entrance. Once inside, march them along the 120-metre reinforced-concrete tunnel that slopes gently into the frozen heart of the vault.

Laughing all the way

By now, the closed-circuit security monitoring system will have relayed live footage to Norwegian and Swedish security departments. Simultaneously, warnings will have alerted the scores of countries involved in the project. Keep the muzzles of your assault rifles to the foreheads of the hostages while holding up a sign to the CCTV, explaining that you will shoot to kill if anything enters the vault. You're no longer talking polar bears. Use the master key to gain access to all three air-locked 10-by-30-metre chambers where the actual seeds are stored. The specimens are kept inside grey aluminium envelopes placed in plastic boxes lined up on metal shelving. Systematically start emptying boxes – each holds four hundred envelopes – upon the chamber floor and grind the seeds under the heels of your boots. Destroy the entire contents of the chamber, move on to the next room and, finally, the third. Mission accomplished.

You realized long ago that there would be no way out, that at this point you would be trapped. Shortly, you will become a martyr, feted across the planet as someone who had the courage and conviction to sacrifice their freedom to ensure others would never have theirs. You will never be forgotten. Of course, the seed bank may, in time, be replenished, but it also might not. You will have proven that disaster can strike anywhere, any time. Nowhere, not even the furthest reaches of the planet, is safe from environmental destruction. The message to world leaders? They can try all they want, but there is no point attempting to protect the planet's future with people like you around.

SEED THE WORLD

WHAT'S THE DAMAGE?

* Domestic flight overshoots Longyearbyen airfield and crashes, slap bang, into mountainside containing vault. No damage is reported to seed bank. **Plausible.**
* Leading botanist suffering profound mid-life crisis loses mind inside seed bank and starts destroying contents, finally locking himself inside vault. **Improbable.**
* By 2020 the seed bank is in constant demand as world experiences a series of crop failures. Vault comes to rescue of US breadbaskets and staves off threats of another potato blight. Millions of lives saved. **Certain.**
* Terrorists hatch plot to break into vault and hold superpowers to ransom. They are arrested at Longyearbyen airport. **Possible.**
* Svalbard Global Seed Bank is left unscathed despite a series of traumatic episodes this century. **Foreseeable.**

Likelihood of Svalbard seed vault being infiltrated: 12%

Blow me

Give me your answer, do

....

AGENDA

* Blame emissions on the Aberdeen Angus
* Beef up your diet
* Chew on a new method of carbon offsetting
* Really milk it

Every plot needs a scapegoat. If you are to mastermind planetary ruination then you'll be looking for someone, or something to take the heat off you. Ideally, you should think about indicting nature or one of her most benign creatures. Something that lolls around in meadows innocently chewing the cud would fit the bill. Something so hapless it allows itself to be reared only to be milked dry, munched up, and spat out. Fodder. Fodder that answers to the name of Daisy.

Now, every time Daisy burps she unleashes a great whoosh of one of the most potent greenhouse gases around: methane. Forget the successes of the oil barons, now that farmed ruminant animals are known to produce up to a quarter of man-made methane emissions, their demonization has begun apace. Yes, these

 BLOW ME 57

conniving creatures are making your emissions worse, expelling a gas with twenty-three times the warming potential of carbon dioxide. You are getting the flak for their flatulence. After you have lovingly nurtured them for succulent steaks, how could they be so traitorous?

Fly me to the moon

Cows really are the must-have scapegoat for anyone serious about f**king with the planet. Blame the bovines and you can carry on peacefully polluting, mining, chopping, and generally just doing what you do best. Efforts to ensure that the cow takes the stick for climate calamity will be wholeheartedly supported by all those engaged in increasing emissions. That's quite a lot of support. The automobile and aviation industries are among those whom you should expect to endorse every attempt to point the finger at the heinous heifer. Already, Michael O'Leary, the head of no-frills airline Ryanair, has tied his colours to the mast. He stated that the massacre of the world's cow population would do more to solve global warming than banning flying ever could. The chief executive of the Dublin-based carrier added that 'eco nuts' should stop fretting about the fastest-growing source of emissions – flying – and concentrate their hot air on livestock farming.

You could go one step further. Take your lead from the Romans and make a sacrifice before undertaking a perilous voyage abroad – a single flatulent cow would fit the bill. After all, each creature emits around 95 kilograms of methane a year, which equates to 2,185 kilograms of carbon dioxide, practically the emissions of two return flights between London and New York, or almost equivalent to eight flights to Paris and back. At any rate, way more than enough to assuage the conscience of passengers trotting up the steps of O'Leary's jets. It would be a simple system. Upon reserving a flight, the electronic booking service notifies a computer database situated in a large warehouse crammed with four-legged herbivores. An electronic pulse activates 10,000-volt nodes attached to each resident. The cow is instantly electrocuted. Holidaymakers are sent the number of their dead cow, which has to be produced alongside their booking reference at passport control. Their consciences are clear and their holidays will be a right laugh because, now they have done their bit for the planet, they fully deserve a drunken weekend break in Honolulu. The several score transport lobbying groups in Brussels should, as a matter of priority, consider pushing for the mandatory killing of cows to become European-wide policy on carbon-offsetting.

Thar she blows

Studies have already been commissioned to explore the potential of butchering these wickedly polluting beasts earlier in life. The less time cows have to belch, the safer the world is. Or so the perceived wisdom goes. On to the crime chart. The belching and farting of a single British cow releases the equivalent of 4,000 grams of carbon dioxide each day. Comparative figures (surely released by the automobile lobby – and if not, why not?) reveal that a four-wheel-drive Land Rover Freelander emits a piffling 3,419 grams on an average day's drive. Further data suggests that the planet's entire herd of 1.5 billion cattle is more lethal to the planet than every car, plane and other form of transport added together. Those who believe global warming is bulls**t finally have their proof.

You've been framed

A 400-page UN report entitled 'Livestock's Long Shadow' was unveiled in November 2007. This little gem is the document with which you can put the cow in the frame. It reveals that farmed animals bred for human consumption cause 18 per cent of global greenhouse-gas emissions. Crikey. If more evidence is required against the cow, merely mention acid rain. When cows pass wind they produce two-thirds of the world's ammonia – one of the principal causes of acid rain. By now, anyone in receipt of such

BLOW ME

information who believes the planet is worth saving may have fled to the nearest field and be already strangling the first cow they chance upon. An honest response, perhaps, but, for your purposes, a naive one. Undeniably, cows are helping advance climate change, but their true value lies in providing a smokescreen for greater, more efficient ways to accelerate global warming. The longer cows are in the dock, the more time there is to pursue airport growth, road-building, economic expansion, and other vital pursuits.

It is with some relief that you'll see even politicians beginning to buy the milky message. Congressman Dana Rohrabacher of California told the climate-change hearing of the House Committee on Science and Technology that previous cycles of global warming had been caused by 'dinosaur flatulence'. Cows, the interpretation went, would bring about the same result. In January 2007, the Liberal Democrats, eager to prove their climate-change credentials, were fast off the mark in identifying the scale of the threat. 'Flatulent livestock emitting methane are beyond a joke,' declared Chris Huhne, who, despite such impeccable green credentials would regrettably, and shortly after, lose his bid for party leadership. Politicians everywhere will soon join in the free-for-all as it becomes evident that the planet's future is irrevocably linked to the mooing gasbags roaming nearby. Daisy and her playmates will be, terminally, put out to pasture.

Petrified by the crisis unfolding in its rural heartlands, the government has already ordered a new research programme, enlisting the help of experts at the Institute of Grassland and Environmental Research in Aberystwyth to discover how this environmental criminal can be tamed. Experts realize that the global-warming gas is produced in cows' guts, but 'how' is a matter of debate, and agreement is scarce, except that this is serious. One experiment involves attaching bags to both ends of a cow to catch flatulence as it is released. Feeding cattle garlic has also been examined. A new type of grass is being trialled. Researchers from the distinguished Well Cow project have implanted sensors to monitor the insides of a cow's stomach. Cattle-lytic converters, it can be assumed, are in development.

A flatulence tax might be called for, but this would only provoke formidable-sounding opposition from a lobby such as New Zealand's Fight Against Ridiculous Tax (FART), a risk too far. Anyway, it's unlikely that cows would be able to fill in the appropriate paperwork, much less pay over the internet.

Kill the fatted calf

Having fitted up this bovine beast perfectly for the methane miasma heating the planet, attention will also be diverted from the wonderfully wasteful way in which food comes to plates on dinner tables and laps across the developed globe. Cows have been around for a while, but more and more people want to eat them. And more eat burgers produced from cows in faraway fields. The UN report which coruscated cows as the enemy actually included emissions from ferrying meat across the world and the making of fertilizers used to raise crops to feed cows. Research in Japan puts it more neatly; suggesting that a kilogram of meat costs the earth 36 kilograms in global-warming gases. But other people, you joyfully observe, are also missing the point. When the Commons Environmental Audit Committee was conducting an inquiry into energy use and climate change, it asked to interview various transport companies yet omitted to request food-processing companies to appear. The bottom line is that, if the world went vegetarian, fewer cows would be bred and we would be left, bereft, with a dramatic loss of carbon emissions. But don't panic, that's never going to happen. That would require an ethical switch, and one of the most reassuring, inescapable factors behind any drive to hasten ecological meltdown is that most humans are not ethical. A lot of people, though, like to think they are, and these should be encouraged to eat organic cows. Research indicates that organic beef might be an even better means of destroying the planet because these cattle emit more methane. Of course, you and

BLOW ME

I also belch and fart, occasionally in open fields, but as yet no one is suggesting we should be killed for the sake of saving the planet. A good thing, really; you are more use alive than dead in the pursuit of planetary dysfunctionality.

You now have a scapegoat, or scapecow, extraordinaire. When things start getting heavy – and the finger of blame will no doubt hover near you at some point (you own this book, for goodness' sake!) – bring up the bovine and keep laying the blame 'til the cows come home.

....

WHAT'S THE DAMAGE?

* New figures show Europe's emissions from transport increased dramatically during 2009. The following day, sections of the media report that such rises are dwarfed by hot air from ruminant livestock. **Highly likely.**
* A herd of cows in Derbyshire is found dead after grass is deliberately poisoned by environmental activists. **Plausible.**
* Eco-groups call for the human race to go vegetarian in order to save the planet. Figures released at the same time confirm that meat production has never been greater. **Foreseeable.**
* Following fresh concern over aviation emissions, O'Leary rants that environmentalists should be strung up and shot alongside every 'f**king cow that ever f**king moved'. **Imaginable.**
* Flatulence tax on red meat voted for by European parliament in 2011. **Tenable.**

Likelihood of cow becoming principal climate-change scapegoat by 2015: 47%

Erode to hell

Polar bears show the way

AGENDA

* Melt the icecaps
* Wipe out the bear's necessities
* Save the Inuits
* Re-carpet your house (or igloo)

Four white blobs drift past, bobbing along on the ice-pocked surface of the Beaufort Sea. Each of the scientists does a double-take. In sixteen years of monitoring the Arctic's inimical waters, they have observed the creatures swimming on three hundred and fifty separate occasions. But this time something is very different. Polar bears are adept, powerful swimmers but these don't seem to be making much of an effort. Perhaps they are exhausted and taking a rest. After all, they have to swim further these days because the rapidly melting sea ice means that stretches of open water are continually widening. But on closer inspection, it turns out these aren't just having a breather. No, in actual fact they have breathed their last.

Bear-faced truths

The polar bear, the pesky poster child of the greenies, has blown the whistle on the once-secret melting of the Arctic, bullying the American government into accepting that climate change is for real after years of gutsy denial. And, as ever, there is no shortage of smug scientists queuing up with statistics to fuel the hysteria: a 40 per cent loss in sea-ice thickness in twenty years; Arctic glaciers shrinking by 17 per cent year on year; half the polar icecap having melted away over the last half-century; an area the size of Turkey recently cracking up and sliding into the sea.

More than any other creature on the planet, the polar bear has evolved into the iconic victim of climate change. The Arctic was designated as your private test-bed, an early-warning weather system where you could delight in observing tomorrow's disastrous changes today; a place crucial for inducing dangerous rises in sea level and changes to oceanic currents. But its sanctity is threatened by these photogenic creatures, lolloping about in their yellowy-white coats. You are losing the public-relations war. It's time to put a cap in the ass of this PR elixir for the environmental movement. It's time to poleaxe the polar.

I'm on top of the world, Ma

Your living room looks cosy enough, but it's missing something. What it needs is a large furry white rug, a voluptuous carpet with a polar bear's head still attached. Book a flight to Yellowknife, in the frozen north of Canada, and begin packing. Once there, a connecting flight will take you to the small settlement of Resolute, in the northern territory of Nunavut, a vast wild Arctic wasteland governed by the indigenous Inuit people. Conveniently enough, Inuits are allowed to kill polar bears for subsistence but, more importantly, they hold the right to sell their tags to trophy hunters. Once in Resolute, state your intentions and name your price. You will pay what it takes, though research indicates that £12,500 should suffice. Set off on a dog-sled with a high-powered telescopic rifle and an Inuit tracker to locate your prey. Resolve to save your bullets for the nine-foot males, who can impregnate several females within each breeding season and are vital for the species' breeding momentum. They are twice the size of a female bear; only an idiot could miss.

Your Inuit guide will be unswervingly supportive of your increasing predilection for dead polar bears. He likes your money. While travelling through the snowy wastes, you could discuss your mutual loathing of environmentalists, pouring scorn on the recent comments from Mary Simon, president of Inuit Tapiriit of Canada, who accused greenies' attempts to protect the polar bear of being driven solely by 'political reasons against the Bush administration over greenhouse-gas emissions'. Her statement that, 'as Inuit, we fundamentally disagree with such tactics' is preposterous. You intend to prove her wrong.

Having Nunavut

The Nunavut government calculates bear-hunting quotas solely on reports from locals, who claim that more polar bears are hanging out near their villages these days. Scientists believe this is because melting ice is driving them inland. So you decide to start paying Inuits £500 for every report of increased 'scary' polar bear activity near their family homes. Sources of income in Nunavut Territory are as scarce as trees and there is no doubt that locals will be happy to exaggerate polar bear sightings for cash. You might even consider fabricating a sighting of a polar bear attempting to attack a small child. On the strength of such reported sightings, the Nunavut authorities recently increased hunting quotas for polar bears by as much as 28 per cent. Although a former Nunavut environment minister denied such a link, his successors will be happy for the extra moolah. The government receives £25 for a non-resident hunting licence and another £400 for each polar bear trophy, plus 6 per cent tax. It all adds up. Already, a sizeable bear-trophy market exists among affluent

ERODE TO HELL

American alpha-males. Once the rest of the world gets wind and quotas are increased to meet demand, you feel confident the polar bear will get its comeuppance.

Bearly there

At the time of writing, there are around 25,000 polar bears left in the world. The Nunavut government currently sanctions a quota of six hundred to be shot a year, of which around eighty are sold to foreign trophy hunters who pay up to £18,000 for the privilege. Most are Americans keen to exploit a loophole in the US Marine Mammal Protection Act, allowing trophies to be brought home. Scientific consensus indicates that, between them, global warming and hunting could prompt a 30 per cent reduction in the polar bear population over the next few decades, a projection that suggests these particular poster boys are well aboard the extinction curve.

Global warming is melting the bears' icy migration routes, a journey critical for breeding and for catching seals. Russian bear populations are threatened by poaching. Pollution is causing deformities and reproductive failures to bears in Norway. Tests have found that chemical compounds used in Europe and North America to reduce the flammability of household furnishings can reach the bears' northern habitats and affect their thyroid and sexual glands. A high rate of hermaphroditism has been observed, making it possible that, when gunning down the odd male or two, you may also have killed a couple of females with the same bullet.

You know the drill

Apparently your Big Oil chums are actually banking on global warming. For them it can't get hot enough quick enough. A quarter of the world's remaining oil and gas is locked beneath the impenetrable frozen core of the Arctic, according to a US geological survey. Once, arguments over who owned this vast treasure trove of mineral wealth were academic. But climate change has made access and drilling possible. Ker-ching.

The new Klondike is already underway, the planet's last great colonial land-grab. More oil naturally ensures more spills, thrills and all manner of mishaps. Provisional tests have seen support ships urgently summoned to tow melting icebergs from a collision course with exploratory oil rigs. And now you, too, can join in with the new oil rush. All you need is a mini-submarine, a British flag, and no qualms (as if!) about creating the next world war. Once kitted out, chug north beneath the Norwegian Sea until you reach the polar icecap where, adroitly guiding a robotic arm into the inky underworld, you plant a titanium Union Jack. You have claimed the Arctic as sovereign territory. That will cause rather a stir. When Russia did the same in the summer of 2007, it triggered an immediate international diplomatic incident. Kremlin cartographers had noticed a narrow isthmus of continental shelf which extended from Russia to the pole. They were having it. But Canada had noticed the same thing and within a day had ordered new icebreakers and stationed a thousand soldiers in the Arctic. Two military bases were announced for the region. Hours later, Danish scientists urgently set sail to lay claim to their share of the region's wealth. Almost simultaneously, a US coastguard icebreaker headed off to map the seafloor north of Alaska. Ω

The first shots of the Arctic conflict had been fired. Relations between Norway and Russia remain tense. Canada, the US and Denmark could easily be sucked into any war over the Arctic, according to EU foreign-policy diplomats. In fact, European governments have been warned by Brussels to plan for a potentially lethal conflict between Russia and the West over the Arctic's vast mineral resources.

Plan away, you say. When hostilities erupt, it's curtains for the polar pin-up. Once the missiles start flying, who will care about a four-legged hermaphrodite who's forgotten how to swim? In the meantime, fifteen oil companies have appealed to explore US-controlled areas of the Arctic. Shell has spent more than £20 million on leases to explore for fuel in the waters where tired

ERODE TO HELL

polar bears are drowning. In Alaska, the US government wants to extract fifteen billion barrels of crude oil from the frozen Chukchi Sea, one of the bears' last habitats, whose treacherous waters provided the iceberg that sank the *Titanic* in 1912. Soon there will be no more icebergs, and there will be no more polar bears – but Godspeed an environmental catastrophe of titanic proportions.

WHAT'S THE DAMAGE?

* Full-scale war declared between EU and Kremlin after Russian warship turns away team of Norwegian scientists from polar circle. **Possible.**
* Investigation into uncorroborated reports of children mauled by polar bears. Inuit authorities say enough is enough and order bear cull. Months later, reports found to be false. **Plausible.**
* As Arctic oil wars escalate, a rig is sabotaged in the Beaufort Sea, causing an environmental catastrophe three times as bad as BP's Prudhoe Bay spill in Alaska, which leached 250,000 gallons of crude oil into a sensitive area. **Predictable.**
* US and Canadian governments come under fierce pressure to stop trophy hunting of polar bears after undercover investigation exposes the trade. **Highly likely.**
* Number of Inuits unable to feed themselves increases. A spokesman blames it on 'blinkered' environmentalists for valuing polar bears above human life. **Probable.**

Likelihood of polar bear population reaching unsustainable levels by 2020: 57%

Seal you later

Fur goodness' sake

AGENDA

* Pelt it out
* Keep 'em culling
* Seal their fate

Once, it was just fashionistas made to feel guilty for the deaths of these wide-eyed creatures but, nowadays, everyone is being told to repent for the deaths of seals. Scientists have monitored the ice on the Southern Gulf of St Lawrence, Canada, and their observations show that it is thinning, cracking, and splintering like crazy paving. The unstable ice means that seals keep sliding off and drowning – and you are being blamed. The crime you are accused of is a (rather unlikely) herculean effort to assist global warming, which is melting the ice on which these seals live. Preposterous. It seems you are being unfairly bashed over the head with the limp, bloody bodies of these cute creatures, despite the fact that the waning sea ice might simply be part of a natural progression, dominated, locals believe, by ten-year weather cycles.

For cod's sake

Seal numbers off the eastern coast of Canada are estimated at just under six million, which is in fact almost three times what they were when culling began in the Seventies. In light of such numbers, Canada refuses to outlaw the seal culls which take place on the ice every spring. Authorities rationally argue that they would only consider the seal population to be in serious danger if it dipped below 1.8 million. This leaves a surplus of a few million seals, which could be readily killed without prompting any action or causing any discernible difference to the planet.

These seals are being accused of a crime against the human being, their greedy consumption of cod ensuring that innocent families starve. Canadian fishermen used to earn their living by harvesting cod, but this is no longer possible. Don't let anyone tell you it's because they have over-fished the seas beyond a sustainable point. No, these pesky seals with their insatiable appetite for fish are ingesting the entire food supply. And if that does not justify the greatest cull ever seen on the face of the planet, then little can. If you were a fisherman, you'd kill for a hot dinner.

Set back from the litter-strewn canyon that is the 'A1 North' is a small business that looks like any other lining this windblown artery of north London. Here, in Archway, is the nondescript UK headquarters of Alaska Brokerage International Limited, a major fur dealership and a specialist in the organized trade of seal skins and furs. Thankfully, the killing of seals for fur and the trade in pelts is not illegal and a reasonable proportion of the company's trade relies on the annual Canadian seal hunts. After the skins are collected from the ice, they are taken to a plant in Newfoundland, from where they are dispatched back to Tromsoe, Norway, for full processing. From there, deals are brokered by external parties such as AB International for onward sale and export.

AB International is run by Peter Bartfeld who brokers deals for Canadian seal skins to be exported worldwide. Undercover reporters have found that Bartfeld's company is the main European agent for GC Rieber Skinn, the world's biggest supplier of dressed seal skin and fur to garment manufacturers. This Norwegian company, a subsidiary of the giant GC Rieber group, which also has interests in shipping, real estate, and minerals, purchases an estimated 150,000 to 200,000 Canadian seal skins a year, as well as skins from similar hunts in Norway and Russia.

Some countries, including Belgium, the Netherlands and the US, have prohibited the import of Canadian seal fur and skins because of concerns over welfare and conservation. Other nations, including Germany, Italy, and Austria are currently considering a ban. The UK government, broadly critical of the seal hunts, has tried to distance itself from the debate over seal products by stating that the UK is not directly involved in this trade. But the existence of the Bartfeld empire, right here in the capital, proves that this claim is untrue. These are men who can help ensure that it's Canadian seal pelts all round.

AB International, believed to be worth in excess of £1.5 million, rightfully claim that their seal skins are the best in the world. Undercover reporters posing as fashion buyers have met representatives of the company and were offered the pelts of 'beaters', young harp seals which cannot yet swim properly. This is perfectly legal, and the younger animals are considered to yield the best-quality furs, with top-grade pelts commanding around £11 each, while lower-quality specimens fetch as little as £4.

The fur industry has recently been going out of its way to increase popularity and sales and, according to campaigners, international fashion houses such as Prada and Versace are among those who've used seal fur and skin in new collections. In the UK, seal products are most widely used in the manufacture of sporrans, part of the traditional Scottish costume.

SEAL YOU LATER

Join the club

Before entering into business arrangements with the Bartfelds you should set down some ground rules. First, their hunters must only use a sharpened club or a traditional hakapik – a wooden bat with a spike at the end like a nail – to kill the seals. No shooting. They'll probably agree, since nothing ruins an £11 seal pelt more profoundly than a bloody great bullet hole.

Secondly, ask Bartfeld to contact his men on the ground and demand that they ignore the tiresome new regulations which were introduced to try and ensure a more humane means of killing. The new edict dictates that, after clubbing or shooting, hunters must check a seal's eyes to make sure it is dead. If it isn't, the animal's arteries have to be slashed. An utter waste when you consider that time out on the blood-stained floes is money. Ask for a clause in your contract which states that any reports of seals being killed humanely will result in the deal being called off. Tell Bartfeld that you will pay double for pelts entirely smothered in crusted blood.

Thirdly, there is to be no close-up footage from animal campaigners. This cull must not be broadcast to the public. The last thing anyone needs to see is more shots of defenceless seals getting walloped to death by burly chaps. Pictures of ice floes becoming criss-crossed with ribbons of scarlet as bodies are dragged home are another no-no. If there is any filming to be done, then it is for you and your friends to watch in the comfort and privacy of your own home.

On Bartfeld's behalf, write to the Canadian Fisheries Department thanking the Canadian people, yet again, for allowing the cull to go ahead. The total quota for the 2008 hunt was 275,000 seals, a reasonably impressive increase on the previous year but still not high enough. Since 2004, around 1.2 million seals have been culled – you can't help but feel this would be a conservative total for one season alone. You have heard the department is fuming, claiming that the debate is being unfairly influenced by the 'emotional rhetoric' of animal lovers. Your letter offers sympathy and understanding and you promise to register your displeasure with the British government for their puerile moral outrage and transparent pandering to 'unfounded and unhelpful' hysteria.

Pest control
You will probably also want to encourage Bartfeld to expand his operations. Seal oil contains decent levels of omega 3 and, in today's health-obsessed world, such a boon cannot be overlooked lightly. And don't forget the creature's penis, a delicacy in the Far East, where they grind it down and knock it back with wine as an aphrodisiac. You can also take action closer to home. Somewhat audaciously, British seals are also eating cod. And our northern fishermen, like their Canadian counterparts, have to feed their families somehow. The North Sea population of 120,000 grey seals have started, according to reports, to eat four times more cod now than they did twenty years ago. Such gluttony must be punished. You may have to take matters into your own hands and travel to the northern tip of Scotland. There, borrow a gun from a supportive farmer – the general rule is that fishermen and farmers get along everywhere – and do the job yourself. Follow the imperious lead of the mystery gunman who, two years ago, shot a whole school of seals at point-blank range on a rocky Orkney beach. The mystery killer took particular care to target only pregnant seals. Superstition says that a person who kills a seal will forfeit his soul to the animal and is destined to return as a seal in the next life. Ignore all that crap. You've got bigger fish to fry.

It is estimated that five thousand seals are killed every year on the remote tips of the British Isles by fishermen and lobster creelmen. Rarely does this make the headlines; no one seems to mind. Here, seals are viewed as pests; the police never investigate. And the Conservation of Seals Act 1970 gives you, in effect, carte blanche to massacre these little pups with impunity. A delightfully worded get-out clause sanctions the killing of any seal swimming 'in the vicinity' of fishing nets or tackle. Thankfully, someone chose

SEAL YOU LATER

not to include a definition of 'vicinity' in the act. For your purposes it could mean a few metres, or any area within the borders of the same country. Bring some friends and make a weekend of it. You'll be doing the locals a favour even if you're only in it for the kicks.

WHAT'S THE DAMAGE?

* Canadian government announces vastly increased seal-cull quotas for 2011 - 510,000 in a single season - after new figures show sharp fall in the region's cod stocks. **Possible.**
* Seal fur makes an unexpected return to the catwalk. Despite widespread protests, Bartfeld's operation booms accordingly. **Slightly plausible.**
* A fur enthusiast is forced to watch seal pups being clubbed to death for a new television series exposing the reality behind the luxury-goods trade. **Likely.**
* Another mystery gunman single-handedly tries to reduce Scotland's grey seal stocks. **Likely.**
* AB International announce record profits in April 2010. Seal pelts become big business. **Probable.**

Likelihood of seals being culled in increasingly large numbers by 2015: 77%

A whale tragedy 11

Having a whale of a time

AGENDA
* Make some noise
* Scramble whale sonar
* Blow 'em out of the water

Imagine a light so ferocious it would scorch your retinas. Blinded and confused, you stumble forward in complete disorientation. For a whale, noise of a certain frequency has the same effect as a dazzling light aimed directly into your eyes. Hearing is the whale's most developed sense, and its spatial awareness is governed by noise. Intense sounds inflict on them the equivalent of blindness. A particularly tremendous noise would probably burst their brain. Here's a chance for you to make yourself heard. To trigger a sonic boom so intense, so traumatizing and befuddling that they would rather hurl themselves to the rocks than face the pain.

A WHALE TRAGEDY

Boom boom, the mighty fall

The Ministry of Defence may have created just such a sound. This recent development may provide you with a most gratifying means of eliminating the whale. And this is some sound system. In secret, the MoD has created one of the loudest underwater systems devised by man, capable of disseminating a sonic boom so thunderous experts predict it could rupture the brains of whales hundreds of miles away. Mammals within range have reportedly been found with blood seeping from their ears. By flooding the seven seas with these sweet melodies you can drive the planet's largest creature to utter madness. In the murky underwater world, where the whale relies completely on sonar for navigation and finding food, this promises to be a satisfyingly apocalyptic weapon. The ancient migration routes of such creatures will be scrambled in the confusion that follows. Countless carcasses will be surrendered to the sea, and nobody will be any the wiser. In your capable hands, this weapon guarantees a stirringly symbolic victory that will strike right at the Volkswagen-Beetle-sized heart of one of nature's greatest creations. That something so prehistoric can be felled by the latest in sound systems aptly demonstrates the triumph of man over the natural world.

Sonar so far

On 14th January 2006 the British public, and perhaps the world, was enraptured by a queer creature which had got lost in the Thames. A northern bottlenose whale, a creature normally at home in the turbulent depths of the North Atlantic, had somehow arrived in the Big Smoke. As his struggles unfolded on live television, the cameras panned out and followed Willy bobbing pathetically past the Houses of Parliament. There, just beyond the nineteenth-century edifice, were the rooms in which Whitehall officials had authorized the development of the very equipment that may have driven Willy into a shallow waterway in the capital. There, almost five years previously, the MoD signed off the secret sonar weapon that some believe may have disorientated the poor whale. Those dastardly enemy submarines had been getting harder to detect and so, not unreasonably, the idea was to build a better detection system. The MoD was impressed with designs for an acoustic version of the atomic bomb and a company called Thales Underwater Systems was given the contract. Sonar 2087 was developed behind closed doors, with details of its strength classified under the argument of 'vital defence capability'.

Early trials made no mention of whales or anything to do with sea life. Then, out of the blue, ministers revealed that Sonar 2087 held the 'potential to be harmful to marine mammals'. Interest was suddenly aroused. Where were these tests being carried out? Precisely how harmful would it be? Did Willy's chums get a bit headachey or were their minds being mangled by the arrival of this powerful low-frequency system? Try as you might, facts are exasperatingly difficult to lay hands on and details of the tests' environmental risk assessment impossible to come by. Findings from a series of covert trials off the north coast of Scotland were never publicized. Repeated demands for a full inquiry into the new sonar system were dismissed.

The official MoD website on Sonar 2087 makes no mention of its titillating potential for harm. Nor does it refer to a UN report that concludes that naval sonar systems pose a serious threat to whales. Studies by the Zoological Society of London are similarly ignored. Zoologists there had conducted some quite inspiring studies. They had examined beached whales whose livers were blighted with cavities like aerated chocolate. The creatures had suffered decompression sickness similar to the bends that afflict divers who surface too quickly. Some irresistible force appeared to have driven them to surface so quickly that they were compelled to practically jump from the sea. Something so fabulously heinous had taken place that these whales preferred to beach themselves on land than face another moment in the sea that sustained them.

Sadly, no one knows how many whales have had their brains

A WHALE TRAGEDY

twisted in such a manner. The impatient wait continues for the first case of a stranding directly linked to Sonar 2087 since it became fully operational last year. Conservationists believe most affected whales die mid-ocean and sink to the bottom. Only a tiny minority, they say, are beached ashore. It is the perfect crime. There is no carcass for a post-mortem examination, no corpse for forensic scientists to analyse. No battered, bleeding body for conservationists to take photographs of and scream horror. No one will ever know, not least the blubber-lovers. In the thirteenth century a whale was seen swimming under London Bridge. It was promptly driven ashore and hacked to death. Those days of open carnage are gone. Now that the nanny state is upon you, murder by stealth is the only option.

Noise annoys

A combination of deduction and unanswered questions indicates the unlikely guise of Sonar 2087 as a not-so-silent killer, the latest instrument at your disposal in the plan to wreak carnage on the environment. And causing no little pain. Details of more than thirty whale strandings linked to military sonar have so far come to your attention. Of particular note are those linked to high-frequency American military systems, which include various accounts of whales being found with blood pouring from their eyes. Others describe internal bleeding around the brain and ears. Details of a mass dash of melon-headed whales into shallow water during a US training exercise may also catch your eye. In many ways, sonar might be the equivalent of driving a tank into a field of deer. A few will, unsurprisingly, leap over the fence.

Thankfully there are still no laws curbing underwater noise pollution. Pleas from a parliamentary committee demanding research into the effects of sound on whales fell on deaf, if not bleeding, ears. However, you can take encouragement from tests conducted by the United States on whales during the development of its own low-frequency sonar. A profound reaction was observed:

most of the creatures fled – sharpish. Those particular tests were limited to a maximum of 150 decibels. The Royal Navy were not going to make a similar mistake: our boys grabbed the sound dial and just cranked it up. Their experiments used a sound level of 180 decibels which, in real terms, is 1,000 times louder. Their tests did not uncover any 'significant biological impact'. Some reports indicate that the underwater loudspeakers used to propel the sound of Sonar 2087 can generate 215 decibels, comparable to that of a jet fighter at take-off. The MoD insist their sonic boom is 'whale-friendly' and, to make completely certain, they even station observers to look out for whales before the sonar is used. This, in anyone's book, should be enough to placate the environmental lobby. Memories of Willy convulsing and expiring on a Kent dockside after being dragged from the Thames will remain with you in the years ahead. Don't let the results of the MoD's post mortem – which concluded that she died from 'dehydration and related causes', detract from your resolve. With all this talk of exploding brains, dehydration is a disappointingly non-graphic way to go.

Weapons of mass destruction rarely come cheap. If you are looking to exterminate some of the planet's most treasured creatures, more often than not you must pay big. A couple of Type 23 frigates equipped with Sonar 2087 will set you back a cool £340 million. Even so, such a small fleet might be sufficient to satisfy ambitions of exploding the brains of aspiring Moby Dicks wherever they may be. A single ping can affect animals in the surrounding 1.5 million square miles of water – roughly the size of the Pacific. Ping away. To move forward with this plan you need to get your hands on one of these units. Thales Underwater Systems do not offer their enviable system for commercial sale, and the cost of navy frigates is clearly prohibitive. Plan B, therefore, is to simply rely on the sonar to get on with its job. Regrettably, this leaves you in the highly unsatisfactory position of trusting undercover national-defence interests to fulfil your task, and so you may never really know the true tally of blubbering victims.

A WHALE TRAGEDY

WHAT'S THE DAMAGE?

* Sonar 2087 conducted in secret off the Hebrides. Scores of pilot whales rock up on coastlines in rather poor shape. MoD reveals it is a 'world leader' in responsible sonar use. **Predictable.**
* Laws to regulate underwater sound levels agreed by international community in 2013. **Never.**
* Defence bods design Supersonic 2087. Terribly effective at detecting enemy subs, classified tests reveal it can blow a whale's head off from several thousand metres. **Anticipated.**
* Another mysterious sequence of whale strandings in Thames blamed on military sonar. MoD denies everything. **Plausible.**
* Use of sonar by military suspended then banned after new research and lawsuit brought by conservationists. **Likely.**

Likelihood of whales being extinct by 2015: 34%

Water shame

Moratorium no more

AGENDA

* Hand out bribes
* Idolize the Japanese
* Shell out for new oil
* Scupper the blubber

As if sonar wasn't enough, here is another surefire method to blow the whale right out of the ocean. Perhaps you think sonar is a little lacking in subtlety, the reverberations and potential repercussions just too loud for you. There is another way. For those concerned with species desecration, never undervalue governmental corruption. A well-established and internationally recognized system not only facilitates the large-scale slaughter of whales but also allows it to be carried out above board. All it requires is a touch of light vote-rigging and, well, job done. As countless elections have proved, few things are more straightforward.

Ballot-aid

Every summer the International Whaling Commission meets to chat about whether they should restart whale hunting. Everybody gets the chance to have their views heard. Some want whales slaughtered with decent haste; the more eccentric seem to prefer their survival. Eventually, the seventy-nine countries vote on the issue and the fate of these warm-blooded blubber barrels is decided. In order to end the exasperating prohibitions of the 1986 moratorium on whale hunting, introduced after many species edged close to extinction because of concerted over-hunting, a 75 per cent majority is required. In these eco-sensitive times, that's a dauntingly high hurdle. The good news is that you can buy votes. Your opponents refer to it as bribery. The more enlightened among you view the system as an open door through which to usher cetacean genocide.

Japan is a past master at manipulating the IWC, and from the bright minds of the Orient you must master how to achieve the legitimate slaughter of the world's whale population. A blend of cunning, determination and solid inducements is required and, from such a pot pourri of talent and gifts, Japan has managed to execute a feat that, in June 2006, left the environmental movement pondering one of its greatest reverses. The news that the pro-whaling block had secured a 51 per cent majority to resume massed hunting kick-started celebrations from Osaka to Ormskirk, an anniversary still honoured by those who stand to benefit. But while the result was totemic, the real reason, as you have since discovered, was rather more prosaic. Japan simply touted huge foreign-aid deals to countries who would back their lust for killing whales. Japan was unashamed of its match-winning tactics. Days after the triumph, a written reply from the Japanese minister for 'aid packages' admitted that his department had given the Caribbean country of St Kitts and Nevis – which had hosted the conference – £2.58 million, although it shrewdly denied any involvement in vote-rigging. Nicaragua was given £8.5 million and the Pacific island state of Palau around half that. In the period after receiving the money, cynics noted that all had developed a sudden, strange desire to catch whales. Both landlocked Laos and Mongolia also developed a hitherto unknown interest in whaling. Sited little closer than 1,000 miles to the nearest ocean, the latter is arguably the dustiest, driest place on the planet. Unfamiliarity had clearly bred contempt, or perhaps you had prematurely dismissed tales of the great Gobi Dick. The Solomon Islands were recipients of cash from Tokyo's venerable Institute of Cetacean Research, although the Japanese government denied any link. With the environmental rewards so undeniably great, Japan doesn't hesitate to opt for the old brown-paper envelope when needs must. Togo turned up late to one conference with its £5,000 membership fee in ready cash.

All these payouts warn that your attempts to restart the wholesale murder of whales will not come cheap. Japan has distributed at least £320 million over the last twelve years. Still, there is a chance you might not have to dig too deep into your pockets. Intelligence indicates that Japan will continue investing until it reaches the magic 75 per cent mark. Once there, it is all over for the whale. Two million were killed in the southern hemisphere alone last century, before the moratorium came into effect.

Currently, the global whale population is more than a million, a number that would be best vanquished in the space of twelve months. Any voting victory could be crushed as little as a year later, and you should not expect a second chance. To achieve this figure, you'll need to cajole the entire fleets of Norway, Iceland and, of course, Japan, the leader of whose fisheries agency once described the minke whale as the 'cockroach of the ocean'. As for the Norwegians, well, they really need no encouragement. Norway, with admirable verve, simply ignores the moratorium, killing more than 25,000 minke whales since it was introduced. Japan might yet just say sod it: its whaling fleet recently set sail for the Antarctic in pursuit of the biggest single whaling slaughter since commercial whaling was banned twenty years earlier. It has a sound argument,

 **WATER SHAME**

which must be propagated as widely as possible. Japan claims that whales eat large quantities of sought-after fish, which is like saying people who eat over-fished cod should be harpooned rather than applauded. On its website, the Japan Whaling Association argues that asking its people not to eat whales is culturally reprehensible, equivalent to 'the English being asked to go without fish and chips'. Quite.

Not so grey days

The principal site of the anticipated whale slaughter lies 400 miles off the north coast of Japan, where whalers and the country's distinguished fisheries department are not the only ones out to destroy this creature. On the island of Sakhalin, one of the world's largest and most controversial oil and gas schemes is about to kill off the Western Grey Whale, a sitting duck if ever there was on the salubrious extinction curve. The Sakhalin II project, a scheme to extract fossil fuels from beneath the freezing seas off Russia's north-east Pacific coast, will cause all sorts of intractable problems for the Western Grey. Oilmen and their equipment traverse slap-bang across the feeding ground of the ever-so-sensitive Western Grey, of which only 120 remain, including just twenty-three breeding females. Environmentalists estimate that the loss of one female per year would lead to extinction. Thank, then, your lucky stars for the involvement of trusted accomplice oil-giant Shell, who are refusing to move their proposed second oil-drilling platform, currently planned to sit adjacent to the favoured feeding area of Western Grey females in the shallows off Sakhalin's north-east coast. The platform construction works are causing the wimpy whales to abandon their only known feeding area, with reports of 'skinny' or emaciated whales among those that still hang around. Shell claim to have already responded to environmental concerns by moving the offshore pipeline. They might want to prepare for even more wearying protests from the greens now that the pipeline has been re-routed through the breeding grounds of many rare birds, including Steller's sea eagle.

Rarely do Shell let you down on such matters, with consistent failure to provide adequate information on noise levels, future construction plans or to draft oil-spill-response plans merely reaffirming your confidence that they mean business. But you must also take some credit. For years you have unknowingly backed this plan – in fact, you were even earmarked to help fund it. Britain's Export Credits Guarantee Department (ECGD), the body using taxpayers' money to underwrite industry overseas, agreed in principle to £500 million for funding the easy extinction of the Western Grey, even managing to avoid carrying out a full environmental assessment of its impact. A Freedom of Information response has confirmed that ECGD's promise of support was made before it had assessed impacts arising from the project, a move that jars with its own policies. The way these guys operate is eye-opening. Not surprisingly, for years the ECGD, whose governors include Gordon Brown, refused to confirm such an honourable tactic.

Now, though, Sakhalin Energy, a conglomerate of oil companies led by Shell, has been forced to withdraw its request for government backing for the vitally important oil and gas project. It has been bedevilled by cost overruns, with its total outlay doubling to around £10 million. You are still confident that fresh requests will be made. You plan to write to the British government expressing your support and asking if there is any way you can help. You might even contact Shell direct and offer what cash you can. Sell your house, flog your body, spread the truth that oil is precious and that Sakhalin is essential for future generations. Although not future whale generations.

WATER SHAME

WHAT'S THE DAMAGE?

* Japan's cultural minister serves up whale meat during high-level talks in Tokyo. Britain's prime minister filmed chewing raw flesh with thumbs-up to camera. Claims later that he mistook it for sushi. **Tenable.**
* Landlocked countries from Bhutan to Burkina Faso suddenly realize they have a soft spot for whale hunting. At 2011 IWC meeting, Japan secures 77 per cent of vote. In ensuing row, Britain demands a Japanese trade boycott. Whaling commission disintegrates. **Possible.**
* Unable to seize a majority at the IWC, Japan rescinds membership and goes it alone by organizing series of massive whale-hunting forays. Norway increases whale catch. **Inevitable.**
* Shell pipes burst mysteriously off Sakhalin, vast slicks swamp Western Grey feeding area. **Plausible.**
* Sakhalin project runs into serious financial difficulties, European taxpayer unknowingly wades to the rescue. **Probable.**

Likelihood of global whale population being extinct by 2015: 12% [for the Western Grey whale: **87%**]

One helluva fungi

Break the mould

AGENDA

* Molest the mould
* Cut off plants' life support
* Give up gardening for good
* Pave over the planet

Small in size and mouldy at heart, the mycorrhizal fungus has emerged as one of your most fearsome adversaries. Without it, experts believe, the planet's ecosystem would collapse. Some say it is the biological cornerstone of plant life on earth, nurturing 95 per cent of plants. It colonizes roots to supply the plant with nutrients and water from the soil, receiving carbon in exchange. Entangled just beneath the surface of the soil, the mycorrhizal resemble a massive mesh of pipework; miles of minuscule tubes, each white strand just a few hundredths of a millimetre wide and invisible to the naked eye. In time, the fungi connect all surrounding trees and plants, creating a symbiotic system for nutrients to be shared.

These are the friendly fungi, the all-round good guys of the planet. Their promiscuous ethos of share and share alike is earning them a reputation. Yet they remain valued and respected by scientists and botanists everywhere. Sickeningly, plants who have hooked up with mycorrhizal are far better equipped to deal with drought, disease, and the attendant turmoil of climate change. If plants get stressed out in your new age of global warming – and you feel warmly confident that they will – these fungi will keep them alive. In poor, denuded soils, they provide a vital lifeline.

The global massacre of the mycorrhizal is essential but, be warned: it has been knocking around for 500 million years and has learnt a few evolutionary tricks. Its prehistoric mutual arrangement with plants allowed both to survive in the days when the planet had crumby soil, was a bit on the warm side, and suffered bouts of extreme temperatures. Global warming will test the supportive role of the mycorrhizal to the absolute limit – eventually the world might have nothing to rely on but this willing fungus. This time, there must be no fall-back. You must mutilate the mushroom.

Put the fun in fungicide

No one said it would be easy. Killing the mycorrhizal fungi is a daunting project, and generous reserves of imagination, dedication, and patience are required. Guaranteed, you will shortly hate these fungi as much as you do the toadstools who campaign to keep polar bears alive. You propose a series of small-scale experiments to supply the blueprint for its eventual extermination. Industrial fungicide should do the trick. Slowly the soil will become brittle and your mouldy nemesis will die. But the problem is that you will need an awful lot of the stuff; it is the sheer ubiquity of mycorrhizal that is so irritating. And fungicide can be damn expensive if you plan to annihilate so prevalent an enemy. A single teaspoon of soil can, for instance, contain 20 metres of coiled-up fungal filaments. Stretch the roots of a five-hundred-year-old beech tree and you will measure around 30 miles of root system. Stretch the same tree's mangle of mycorrhizal fungi, lay it out end to end, and it would measure 32,000 miles, enough to wrap around the planet. Its elimination will thus require manpower comparable to that of a small army. Recruiting volunteers may prove fairly straightforward, however; most people have never heard of mycorrhizal and, as a species, it is unlikely to garner much sentimental support. The word 'fungus' hardly inspires a passion. Start by targeting gardeners, who already tend to view fungi with an irrational hatred.

At this stage, you believe that a biological weapon – an efficiently voracious mycorrhizal-eating virus – is the most likely tool to achieve your aims. You ring around the various government laboratories and ask after the mystery mycorrhizal massacring agent but, incredulously, the boffins have neglected to pursue such a vital brief. You do learn, however, that the mass use of agricultural chemical MPK fertilizer successfully prevents the fungi from getting a decent foothold in soils and, as a welcome side-effect, has been associated with some of the best excesses of industrial farming. It results in soil becoming denuded, in need of constant soaking to ensure productivity. But yet again, this scheme will prove too costly for the quantities necessary and, of course, the downside of using fertilizer to kill the fungi is that it encourages other plants to grow. And you're not a fan of plants. They take too much looking after and drop leaves all over your tidy garden. You like bare earth; you like cracked, withered dusty topsoil, so frail that even the hardiest cacti would rather move on than hang around eking a desultory lifestyle from such a soil.

The idea of a scorched-earth policy excites you. You locate an acre of land and hack down every last plant. Then, using a magnifying glass, and submitting to your obsessive streak, you yank up every spot of green. It is hellishly time-consuming, back-breaking work and even you, ever the eternal optimist, are

ONE HELLUVA FUNGI

somewhat daunted by having to ensure that every scrap of plant life is removed and never allowed to grow again for ever more, amen. You may receive considerable goodwill from your supporters, but this prospect might exhaust even their reservoir of faith: it would involve everyone jacking in their jobs and devoting their existence to the daily scrubbing of the planet's surface to ensure that it is clean of plants.

Sully the soil

Alternatively, you could always just concrete over the entire planet, make the world one great patio stretching from your back garden. It would be rather useful for people to park their cars and land their planes on when they pop along to one of your 'end of the world' bashes. A simpler option might be just to get out the old rotavator and churn as much of Europe as possible before someone asks what you are doing. Mycorrhizal absolutely detests disturbed ground, so the invitation to interfere is implicit. Or, consider those formal gardens so favoured in the Seventies, the ones with neat herbaceous borders and large bland gaps between a couple of rosebuds. Again, the fungi – and not just for aesthetic reasons – cannot stand this. The downside here is that you've created a garden that looks like your great aunt Agatha's and, even in the canon of ecocide, that might still be too high a price to pay.

It is hard to accept, but increasingly tenable, that the mycorrhizal might outwit you. And yet the more you hear about it, the more you know that its survival cannot be countenanced. Its temerity is outstanding; it even manages to protect plant roots from pathogenic fungi such as nematodes and bacteria. As usual, you are operating in a finite window of opportunity. Already, a growing consensus among the more dislikeable elements of the gardening set is for a return to natural, organic systems using mycorrhizal to facilitate fertility. Recent warnings reveal that reserves of phosphate, an indispensable but rapidly depleting fertilizer for farming, will remain economically viable for no more than a matter of decades if current rates of consumption continue. International experts claim that without phosphate there will be no agriculture, no food to sustain humanity. But, for no good reason, they ignore the existence of mycorrhizal which, although less productive, may rise like a boil on the face of the planet to help sustain crops and plants under threat.

You will only be content when the world runs out of phosphate and you have killed off the earth's friendliest fungi – then everyone really *is* in the doghouse. In the meantime, you must keep searching for the silver bullet; you must up the hunt for the ultimate biological weapon. Eventually, you could even consider going nuclear, bombarding the planet with atomic bunker-busters once designed for flushing out al-Qaeda figurehead Osama bin Laden but now to be engaged in flushing out a far more slippery foe. Employ napalm. Organize deep underground minefields. Far-reaching acid baths. Anything is in and nothing is out in the holy war against the mycorrhizal.

WHAT'S THE DAMAGE?

* Previously unknown voracious virus begins sweeping planet and devouring mycorrhizal at a rate that stuns biologists. **Improbable.**
* As effects of climate change take hold, the fungus proves it is not as effective as predicted. Crops wiped out across the plant. **Maybe.**
* Importance of mycorrhizal begins to be more widely realized and gradually becomes purchase of choice for gardeners everywhere as consensus shifts from man-made fertilizers. **Highly likely.**
* Scientists pronounce mycorrhizal as critically important and UN demands it the first fungus to be declared a protected species. **Possible.**
* Phosphate reserves expire sooner than expected. Emergency summit held to discuss replacing world agricultural systems with a fungus that does not harm the planet. **Plausible.**

Likelihood of mycorrhizal being eradicated by 2015: 4%

Going ape 14

It's all monkey business

AGENDA

* Get Attenborough off our screens
* Take sides with rebels
* Big business behind you
* Stop aping around

For once, David Attenborough seems lost for words. As he kneels beside the mountain gorillas of central Africa during an historic encounter for the 1979 BBC *Life on Earth* series, the veteran broadcaster is overcome with awe. 'There is more meaning and mutual understanding in exchanging a glance with a gorilla than any other animal I know,' he whispers. The primates crouched before him have DNA which is 97.7 per cent identical to that of the human being. As Attenborough's cameramen gently follow the family of gorillas feeding, you might as well be watching yourself through the lens.

They are one of the most symbolic creatures on earth. No other species has won more love and curiosity than the mountain gorilla; the ultimate international icon of all that is endangered. Nothing could rival the extinction of humanity's closest cousin as confirmation of man's power to dramatically alter the face of the planet. In one stroke, our evolutionary bridge would be obliterated for ever. Darwinists will gasp in horror. Conservationists will lament the loss of a keystone species. You will have seen off your nearest relative in the relentless progression to a mighty goal.

Recent events play a helping hand. Already Attenborough's apes are staring into the abyss. They are innocents at the epicentre of the deadliest war since Hitler's armies marched across Europe. The Democratic Republic of Congo is not only home to more mountain gorillas than any other country, but its jungles neatly double as the battleground for one of the world's most lethal conflicts. The official ceasefire of 2005 is doing little to stem the inevitable reprisals of a struggle that has claimed four million human lives. And it is the fug of this war that will play host to the most audacious extinction in modern history.

Gorilla warfare

More than likely, Karema stretched out a welcoming hand as the gunmen approached. Fully habituated by tourism, the solitary Silverback trusted implicitly the figures marching closer through the jungle mist. Karema was held down and hacked apart with machetes. His heart, arms and legs were eaten by rebel soldiers for breakfast one morning in January 2007. Renowned for his exceptionally calm personality, the huge gorilla was the archetypal gentle giant and would not have fought back. His remains were found two days later, smothered in human excrement at the foot of a pit latrine. News soon spread that another Silverback, one of the males crucial to controlling the species breeding pattern, had also been scoffed by the same gang of anti-government rebels. Gorilla meat, the soldiers swore, made them courageous killers.

Just two isolated, tantalizingly vulnerable pockets of gorillas survive. Around 380 live in Africa's oldest national park, the Democratic Republic of Congo's Virunga Mountains, whose looming volcanoes and pristine rainforests, popularized by the film *Gorillas in the Mist*, are known as the gorilla sector. Another 320 gorillas are to be found in the adjoining jungles of Uganda's Bwindi Impenetrable National Park. Also buried deep inside these forests are men with the impunity, weaponry and, allegedly, the moral compass required to vanquish the creatures that so astonished Attenborough.

Karema is believed to have died at the hands of those under the command of Laurent Nkunda, a gaunt, university-educated 38-year-old whose rebel armies hide among the Congolese southern forests. Fate has thrown your man Nkunda and the world's few remaining mountain gorillas together. Commander for rebel forces known as Rally for Congolese Democracy-Goma (RCD-Goma), Nkunda is wanted for appalling war crimes including summary executions and mass rape. Violent oppression is the group's *modus operandi*. Murdering the odd gorilla shouldn't pose too much of a problem for a figure of such calibre. His whereabouts must be known to the United Nations, since a warrant for his arrest was issued more than three years ago, but scant evidence exists to suggest serious efforts have been made to seize him. And a good job too, as Nkunda seems beautifully suited to orchestrating this outrageous operation. Getting him on your side poses the first obstacle but there's an obvious solution. Guerrilla campaigns are not cheap. Nkunda no doubt requires constant funding to finance militia attacks against the Congolese authorities. Procuring his services, therefore, should involve nothing more than a simple monetary transaction. Karema would have fetched around £20,000 as bush meat. An estimated eighty wild Silverbacks are surviving in central Africa. If Nkunda does the maths, as little as £1.5 million might persuade him to see off our remaining ancestors.

GOING APE

Laudable conservation efforts ensured the tiny population of mountain gorillas clung on throughout the civil war but the few, under-funded wardens tasked with protecting them will be easily overcome by Nkunda's well-trained fighters. Karema's remains were found, tragically, near to an abandoned patrol post. Almost a hundred park rangers have been killed by poachers in the past decade. Protecting the mountain gorilla is the deadliest job in world conservation and only the mad or simple of mind need volunteer. Nkunda's militia regularly use Virunga as a base from which to launch incursions into the Bwindi Impenetrable National Park of Uganda, whose similarly modest sprinkling of wardens can also be expected to offer paper-thin resistance. In other words, all surviving Silverbacks are found within the killing range of Nkunda with little or no protection against his vices.

In good company

Actually getting your money to Nkunda in the first place is more fraught. For that, attention must turn closer to home. A number of major British companies have a track-record of trading with men like Nkunda, a policy that helped prolong the political instability that encouraged civil war. Although the UK government donates generously to the Democratic Republic of Congo, it has also turned a blind eye to UK firms' behaviour in the country. One company with longstanding links to Nkunda's RCD-Goma is London-based Afrimex which trades in minerals found in central Africa such as coltan, critical in the production of mobile phones. New evidence reveals that the company made tax payments to RCD-Goma from 1996 onwards, at the same time as the rebels were committing a series of human rights abuses. Afrimex directors say they were not aware of how this money was being spent. All claims are yet to be investigated.

Five years ago a UN inquiry recommended action against a number of companies, including many in the UK, for plundering the Democratic Republic of Congo's natural resources and dealing with criminal and rebel networks. Not one has been sanctioned by the British government. Your safest option, therefore, is to channel funding behind the respectable facade of some of Britain's biggest companies. The government's reluctance to regulate its companies inspires confidence. The conclusions of the influential House of Commons International Development committee said that the failure to satisfactorily conduct inquiries undermined Britain's claims to corporate responsibility. Only one case appears to be still under investigation at the time of writing. These are allegations against Das Air, based near Gatwick airport, and just down the road from the Crawley headquarters of African International Airways. Das Air are accused of flying five million bullets to the eastern stretch of the country, the area in which Virunga is located.

In the same month Nkunda's arrest warrant was issued, world delegates congregated 400 miles from Virunga in the country's capital, Kinshasa. They heard that extinction of the great apes was more than likely by 2010 unless new conservation targets were agreed. Four days later the first comprehensive strategy, the Great Apes Survival Project Partnership (GRASP), was unveiled to international acclaim. Dr Klaus Toepfer, executive director of the UN's environmental agency, said that a minimum of £12.5 million was desperately needed by the close of 2005, an amount he described as 'the equivalent of providing a dying man with bread and water'. Much more, he added, would be needed five years later if the great apes were to remain in the wild. By spring 2007, just £2.5 million had been received. The British government sent an extra £50,000. An internal memo named almost thirty donor companies, three of which were British. The support of governments and businesses was critical, the bloke from the UN had urged. This Toepfer character must surely have known that once you start having to beg companies for help, the game is almost up.

GOING APE

Single-handed slaughter

Maybe, after all, Nkunda is surplus to requirement and the mountain gorillas of central Africa will fade away without giving him the resources to acquire sixty-three brand-new AK47s. Or, come to that, having to delve into the darkest corner of Britain PLC. The gorillas' habitat is shrinking, as chainsaws rip through the base of yet another tree. Even without your funding, guerrillas murdered ten gorillas in central Africa last year. The most cost-effective strategy might be to leave things as they are and save your money for when it is really required.

And while plenty of concern is being voiced, the creatures who made Attenborough lose his tongue for a grateful moment are running short on real friends. Karema adored spending time with humans and smiled in the company of his keepers. But when the gunmen approached through the mist that January morning, Karema was only able to outstretch his right palm. The name 'Karema' means handicapped, a name he was given because, while frolicking in the forests as a youngster, his left hand was ripped from him by a snare. Men plagued his existence from start to end.

WHAT'S THE DAMAGE?

* Nkunda arrested by UN and brought to trial. He is sentenced to life imprisonment for 'perpetrating innumerable, wicked acts against fellow humans'. **Never.**
* In a surprise twist, British government prosecutes companies who have aided and abetted known criminals in the DRC. **Unlikely.**
* Rebels attempt to hold DRC government to ransom by threatening to exterminate all mountain gorillas, a key source of tourism funding. Authorities stand firm. Mountain gorillas are declared extinct after four-day massacre. **Possible.**
* A massive £100m international fund to save the gorilla is unveiled in 2010. Number of wildlife wardens quadruples. **Improbable.**
* Footage of rebel soldiers cooking the limbs of a Silverback in a cauldron of curry appear on YouTube in 2009. One shot shows young soldier eating its heart, the gorilla's giant hand doubling as a bowl. **Imaginable.**

Likelihood of mountain gorilla population being declared unsustainable by 2020: 67%

GOING APE

The Ends of the Earth
How to wreck the land and sea

15
Spruced up

Shiver me timbers

AGENDA

* Set the Siberian forests aflame
* Permafrost no more
* Rupture the lungs of Europe

Stretched out like a humongous tinderbox across the top of the world's greatest landmass lie the forgotten forests of Siberia. The largest unbroken tract of trees on the planet, this is a critical part of the earth's respiratory system. These boreal forests represent a formidable barrier to your plans to despoil the world. Not only do they have the audacity to soak up the immense carbon emissions that drift across Siberia from the industrial powerhouses of Europe, but they also add insult to injury by safely locking away a quarter of the world's methane within the frozen soil beneath.

To accelerate climate change, you must burn these forests, liberating 70 billion tonnes of potent methane from the Siberian permafrost. Climate change has adroitly transformed these trees into a 2,000-mile length of kindling. Once the trees are suffused in flame, there can be no winding back the clock. Siberia's forests are evergreen, their leaves coated in natural oils. They may as well be doused in petrol. Anyone who has watched a Christmas tree turn into a Roman candle will testify to their flammable properties. This vast landscape of spruce, larch, fir and Scots pine is a pyromaniac's playground. The large-scale melting of the permafrost will be irreversible. You can incinerate places overnight, but you cannot re-freeze the world's largest sub-zero peat bog.

The fun of the fire

NASA scientists monitoring satellite images were used to seeing fires rage across the earth, but this was something else. A mosaic of 157 large white splurges, each blaze smothering a section of Russia 27 million acres across. A dark smudge hung across the face of the planet, more than 3,000 miles wide and reaching right down to the Japanese city of Kyoto, where six years previously the international agreement on climate change had been hammered out. Now, these prodigious fires were spewing out huge amounts of greenhouse gases, equal to Europe's total greenhouse-gas-reduction targets under the Kyoto protocol.

11th June 2003 will be remembered as a camp-fire compared to the inferno you will unleash. The time is right to strike. Russian scientists warn that their boreal forests face an unprecedented fire threat. These forests, home to half the evergreen trees on the planet, have become money to burn. Your mission begins on the Siberian border with China, where your accomplices should be hand-picked from the network of adept arsonists already operating within this remote landscape. You are spoilt for choice. A recent single inspection by government officials within the isolated Krasnoyarsk region found a hundred illegal timber processors and 894 timber trucks operating without official paperwork. These workers are accomplished at burning the forest without getting caught. Some have testified to investigators that they report only a third of fires, at most. Those which are reported are deliberately downplayed and companies never admit that a fire has lasted for more than three days. Remember to keep bribe money handy. Police assigned to regulate the trade have been described by undercover investigators as local 'militia' willing to turn their backs for the odd greenback or several. A float of £10,000 in cash will cover payments to fire-starters you meet along the way.

As a decoy to your arsonist activities you might apply for a logging licence. However, permits to log healthy, pristine forest areas are expensive, and in the interests of economy you should avoid paying full rate. The cheapest licences are given for clearing areas of timber scarred by fire. Chinese companies happily pay the same price for fire-ravaged timber as they do for premium timber. This might explain the tendency for timber merchants to become such skilled arsonists. Don't panic about such practices; they are not illegal because the fires can actually make logging easier, sweeping through areas of restrictive foliage and undergrowth before moving on, leaving the trees themselves intact.

Tim-ber!

Having set a satisfying large swathe of south-western Siberia aflame, it's time to move deeper into the dried-out forests. Hook up with the mobile, well-equipped teams of woodcutters that venture for weeks on end into the heart of the taiga. Government documents reveal how these chaps infiltrate the most protected areas, cherry-picking the most handsome trees to flog to the Chinese. Then they cover their tracks by starting very large fires.

You might opt to place an extraordinarily large order for timber. Create a false company, pay upfront and ensure that the order requires enough new trees to cover, say, the size of Britain. Be prepared to pay at least £100,000 as an initial instalment, insisting you will pay for any other trees damaged if the fire gets, ahem, out of control. Target operations from late June to early July, the traditional start of the fire season, when the evergreens are at their

SPRUCED UP

thirstiest. The good news is that the burning season is now longer than ever, with spring arriving several days early in recent years. When the twenty or so fires you finance eventually burn out, leaving a depressingly large swathe of the taiga still intact, you will have long ago liquidated the company and gone underground.

Exporting timber would bring in enough money to fund a fresh round of arson attacks. Illegal exportation is simple because unlimited help is at hand. Documents reveal that militiamen at the district level, members of the Highway Patrol Service, as well as official inspection teams, have all accepted bribes for allowing illegal timber trucks to pass. Bribes range from a bargain £8 to a perfectly reasonable £17, or a job-lot three trucks for fifty quid. The easiest way to get your timber out safely uses the Belogorsk-Norsk-Fevral'sk route and heads through the Svobodnenskii or Shimanovskii districts. Follow the large tyreprints from other illegal timber trucks and proceed towards the two large rail shipment depots in Belogorsk. There, advertising boards are marked with 'timber reception point'. As long as you have paid your dues it should prove a relatively hassle-free journey.

If you researched your trip thoroughly, you'll have found no mention of anyone being charged with illegal logging in Siberia. Arson also doesn't seem to be a high-risk affair and records of police investigation into fires are limited to Siberia's oldest synagogue and a mental-health hospital, with not a word on trees. Don't concern yourself with the threat of inspections. You might find details of 17,000 inquiries by the State Forest Agency and about 4,000 recorded misdemeanours, but a search for subsequent sentences will draw a blank. If in doubt, just bribe someone. Even Moscow officials can be bought according to investigators.

Fly in the face of fighting fire

Once ablaze, the most practical way of taming these fires is to use specialized firefighting aircraft. Moscow designed and built planes for this exact purpose, but in a shrewd if strange decision by the authorities, they have all been leased abroad. At the Krasnoyarsk forest-fire laboratory, deep in the taiga, the frustration is palpable. In this town, where the British Council once staged a Zero Carbon City exhibition, the firefighting team can only watch the surrounding forest burn. Expensive leasing negotiations with countries such as Turkey are required to get back the Russian-made aeroplanes designed to protect the taiga. Outdated equipment, barely capable of snuffing out a candle, is all that stands in your way. Under the old system, forest protection was well funded, but the laboratory recently declared itself 'practically penniless'. Trees have become a bit boring since Soviet times.

Despite everything you've just read, there is no immediate panic to raze the great forests of Siberia. Unlike other time-pressed ways to f**k the planet, you might have the luxury of a relatively relaxed approach, if that's what you prefer. Research at the University of California indicates that the very existence of spruce forests could in fact make global warming worse. Studies in the Rocky Mountains suggest that, in warm and dry conditions, forests give out more carbon than they absorb. During increasing temperatures, fungi and microbes speed up their respiration and release more carbon dioxide into the area than the forests can soak up. Something to bear in mind, if you're not hot about fires.

Like a moth to a flame

Good vibes will be further fanned by the tiny wings of the Siberian moth, which is determinedly helping out in preparing your kindling. This moth devours the needles of the evergreens, causing the trees to die and dry out. About 80 per cent of moth-infested forests will catch fire at some point within the following decade. You can make it 100 per cent.

Meanwhile, a slab of permafrost spanning 250 million acres – an area of France and Germany combined – has started to melt for the first time since it formed 11,000 years ago. Scientists are sending back encouraging observations. In places, the methane

has bubbled to the surface so quickly that it is preventing re-freezing. Soon the great fires will start. Where once a mat of green cloaked the top of the Eurasian continent, there will be a choked, post-apocalyptic landscape of charred stumps and brown bogs. A true vision of the future.

WHAT'S THE DAMAGE?

* After a huge number of fires in June 2009, Russia invests in fleet of firefighting aircraft. **Unlikely.**
* Fires rage out of control for a record seven weeks in 2010. An area the size of Europe is incinerated. Satellite images monitor plumes halfway across the planet. **Certainty.**
* Methane emissions from Siberian forests hit record levels. The region becomes a patchwork of lakes bubbling with the much-loved greenhouse gas. **Not likely.**
* China demands the Kremlin erect a huge ten-metre firewall along the border, with special fire exits for logging companies. **Credible.**
* The Siberian moth enjoys a peculiarly productive period. Half of the entire forest is contaminated and dries out. One morning in 2012 it catches alight. Commentators swear they can actually feel the planet heating up overnight. **Probable.**

Likelihood of fire destroying majority of Siberian forests by 2020: 64%

Con with the wind

You win sun, you lose sun

AGENDA
* Tilt at windmills
* Say no to solar
* Wave goodbye to wind power

Some time back, the leaders of this island were told they could power much of their land by harnessing the power from the waves and wind that surrounded it. After years priding themselves as the 'dirty man of Europe', suddenly they were in grave danger of undoing all their hard toil. Nonetheless, they obediently agreed to Europe's proposals that, by 2020, 15 per cent of energy must come from renewables.

A complete wind-up
But the leaders of this island nation were telling porkies; they never had any intention of giving in to such European guff. The British government deplores the idea of being reliant on renewables. As if

invisibly guided by your masterly hand, it has embarked on a delectable tapestry of obfuscation, delay and untruth of a type that millions of pounds' worth of lobbying could never hope to buy.

It's an ill wind ...

While ministers continued to bang on about thwarting climate change – not surprisingly fooling most of us hapless citizens – they were doing all they could to kill off the embryonic renewables revolution. Like chocolate soldiers marching into the heat of battle, ministers kept insisting that the UK was showing 'leadership' in the area. We were streets ahead of other nations in tackling climate change, the prime minister said.

But as ministers talked up their renewables dream, the scores on the doors revealed that only 2 per cent of the country's energy came from the wind, waves and sun. Of our European neighbours, only Malta and Luxembourg had bothered less. This didn't faze them – ministers continued to extol the merits of clean energy while simultaneously orchestrating their own form of dirty protest. In 2007 the sector began laying off staff due to lack of government support. Grants were slashed, initiatives pulled, schemes revised. People who wanted to 'make a difference' suddenly found it uneconomical. Those who had their eye on a mini turbine for their roof had the wind quite literally taken from their sails when grants were halved. The government had promised tax relief for 'zero-carbon' homes which drew electricity from local renewable sources. Craftily, though, it had made sure that no such energy existed. Of Britain's 21.2 million homes, six applied.

So long, solar

When a high-profile grant was unveiled to develop wave energy, two applicants came forward. Both were rejected. The government, navigating the choppy waters of semantics, would gladly reward development projects, but only those that were fully developed and already operating. You can only applaud. They set aside the impressive figure of 0.01 per cent of gross national product to facilitate Britain's move towards becoming a sustainable society. An internal briefing in the summer of 2007 concluded that there was no chance of going green, since achieving 9 per cent of energy from natural sources was 'challenging'. Civil servants considered asking the EU if they could include nuclear power in the figure, and, since the government had spent some money on renewable energy in Africa, what about including solar power from Nairobi? Ultimately, officials decided to lobby for a more fluid interpretation of European targets, which were unrealistic and lacked credibility. Forget that Germany was hitting the 27 per cent mark; they were simply showing off to embarrass Britain.

A following wind

Yet it is perhaps unfair – if not unwise – to let the government do all your hard graft. Britain is blessed with a charming set of folks up and down the island who'll seemingly do anything to sabotage attempts to save the planet, especially if these attempts involve mild interference with the view from their lounge, or building something futuristic within 7 miles of their rear lawn. Meet the Nimbies, a loose-knit but indomitable fighting force whose ability to attack with irrational fury and peculiar tenacity on issues of outrageous triviality is unrivalled anywhere in the world. With the help of your Nimby friends, you can ensure that wind energy is derailed at a stroke. In fact, blocking turbine planning applications is practically their *raison d'être*. Contact the Country Guardian group, which claims to have halted or postponed 89 per cent of planned wind farms. Their vice-president, Sir Bernard Ingham, has links with nuclear industry and BNFL. If they're too busy blocking new windmills, try the cunningly named Renewable Energy Foundation. Chaired by *Deal or No Deal* host and all-round good guy Noel Edmonds, the REF is well linked to scores of groups around the country who exchange advice on how to undermine the industrialization

CON WITH THE WIND

of the countryside. Heart-warming research reveals that two out of three applications for onshore wind farms are being rejected. Thirty-three have been turned down by Nimby-terrorized local authorities in the past eighteen months. Schemes to provide the equivalent in power to that generated by eight conventional power stations languish in the planning system. How can you fail to salute them?

Tilting at windmills

Despite the fearsome opposition of middle-class homeowners, some offshore wind farms remained determined to deliver. Step forward the military: these wretched windmills interfere with radar systems. Also, do not underestimate the power of leafleting. Keep an eye on planning applications and distribute leaflets to surrounding homes giving overblown descriptions of the plans. 'Imagine the Empire State Building with steel blades longer than a Jumbo Jet ...' – that sort of thing – and compare the wind farm with such outrages to personal freedom as the seatbelt and income tax. Make sure to liken the battle against wind farms to defending the south coast from the Nazis. Arrange pickets against any councillor who supports wind power in their unspoilt village. Nimbies may have largely decamped to the shires, but they will never be as green as their wellies or their Land Rovers. Maximize support from such august organizations as the Royal Society for the Protection of Birds, whose membership offers one million potential recruits for your plan to sabotage clean energy. The RSPB is not overly partial to wind turbines. Write to them in graphic detail about how the blades are mashing up pretty birds in your area and send Photoshopped images of mutilated avian corpses. Claim that turbines can cause brain tumours in humans, documenting a series of mystery illnesses in an unidentified Cornish hamlet since a windmill recently appeared on a nearby hill. Encourage Nimbies to write strongly worded, one-sided letters. It's what they're best at.

Of course, a hotter planet of long droughts and busy beer gardens might also open the door for solar power. Unnervingly, photovoltaic power is getting closer and closer to becoming economical. So far, the government has that dismal scenario covered. When it said it was going all guns blazing for solar power, it began offering grants to persuade homeowners to install solar panels. By all accounts, the take-up was annihilatingly impressive. Around 270 homes, in fact. In Germany, the comparable figure was a smug 130,000. Germany added more than 180 times as much solar power energy on to its electricity grid as the UK that year. But the UK government, satisfied with its own performance, slashed solar grants, ensuring it was just on the cusp of being uneconomical to install. The amount of electricity saved dropped to 2 per cent, a magnificently achieved practical irrelevance.

Power to the people

One way to render these paltry savings absolutely null is to ensure that electricity demands keep growing. More gadgets, another laptop, bigger plasma televisions to keep on standby 24/7. Even if the Dales were covered in windmills, expanding needs would suck up the new energy. It's like running the wrong way up an escalator – and there's always an idiot who'll do that.

Sadly, word is starting to spread about the government's private intention to neuter renewables. Unwisely, the give-away was when they allowed carbon emissions to increase. By the onset of 2008, ministers were in a right pickle. It was time to submit their latest carbon-emission figures. The government is no slouch at 'statistical interpretation', but it is doubtful they could have been more brazen. UK carbon emissions were down by 12 per cent since 1990, they said, to the sound of imported champagne corks popping. That would tell the UN lot to shut it. But those tiresome enough to check the fine print soon found that hundreds of millions of tonnes of greenhouse-gas emissions had vanished from official figures. Those in the numbers department had included

CON WITH THE WIND

carbon credits and subtracted aviation, shipping, overseas trade and tourism from the stats. For the record – and this really calls for a champagne celebration – Britain's carbon output actually rose by 19 per cent over that same period, with everything included. To capitalize on this good vibe, the UK's ambassadorial role in urging other countries to sort out their climate-change policies is profoundly, perfectly, compromised. Rule, Britannia!

WHAT'S THE DAMAGE?

* A rare corncrake is killed by a turbine. The Nimby machine goes into overdrive. Noel Edmonds can't resist mentioning the tragedy on *Deal Or No Deal*. Turbines are torpedoed. **Possible.**
* Investment in wind and wave energy increases to respectable levels. Britain involved in meeting targets set by Brussels shocker. **Impossible.**
* Grants to solar and wave energy slashed further. Windfall taxes added to industry. Entire renewables sector decamps to Germany. Hours later, Gordon Brown announces that Britain is leading the world in climate change. **Probable.**
* Sir Bernard Ingham becomes MP on anti-wind-farm ticket. His manifesto reveals they are 'built by the hands of Satan'. **Plausible.**
* Another leaked memo heaps further humiliation on government. It exposes how civil servants spent £500,000 examining whether it was possible to stop the wind blowing across 'goody two shoes' Germany. Anything to help the UK to catch up. **Imaginable.**

Likelihood of Britain failing to reach European renewables targets by 2015: 94%

Bottom trawling

Don't just scrape the surface

AGENDA

* Score the seabed
* Diminish the fish
* Plunder Davy Jones's locker
* Erode the Queen's Bottom

A giant airship passes over the capital, dragging beneath it a huge net of chain metal. As it moves, everything in its path is scooped up in a crumpled mass of rubble. Finally, the net is hoisted, leaving behind a smooth, featureless canyon running through the heart of London. All that remains of Covent Garden and Leicester Square is a spectacular furrow bereft of life. All in all, a quite impressive feat, based on a fishing system honed in secret beneath the sea.

There, rather than airships, beam trawlers drag gigantic nets weighed down with 10-tonne weights. The areas of seabed they cover are left scorched bare, everything indiscriminately hoovered up, a fishing technique comparable to hunting rabbits with napalm. No one knows how much of the seabed has been desecrated in this manner. On the surface, the sea still evokes its beautiful tranquillity, but in its depths only carnage. Long ago, Britannia ruled the waves, but with her empire in tatters, she now sets her sights on dominating what lies beneath.

Trawling for trouble

It was in the netherworld of the North Sea that scientists first realized something had gone terribly wrong. In one of the most inhospitable sites under British sovereignty, they discovered stunning coral blooms three times the height of man. As the scientists excitedly scoured the underwater images, they came across something else so startling it is hard to imagine their horror. Gouged deep into the seabed were mysterious wounds, each up to 25 miles long. But these weren't formed by nature. No, even here, beneath hundreds of feet of water, man had made his mark.

Having emptied Britain's shallow coastal strip of its once bountiful fish stocks, fishermen had delved deeper into virgin territory: the seabed itself. These spectacular deep-sea scratches, like rugged mountain gorges, were caused by modern fishing equipment. Beam trawlers, indisputably awe-inspiring pieces of marine machinery, are an essential component of your armoury against the planet. Their force is incredible to behold. Underwater cameras have witnessed 'clean' rocks, enormous slabs that have lain undisturbed since the last ice age, scooped out in a single motion. During a typical fortnightly fishing trip, each British trawler scrapes clean more than 440 football pitches of seabed.

If this sounds like an art-form you might be interested in, your first port of call should be one of Europe's largest privately owned fishing-fleets. Based at Newlyn in Cornwall, Stevenson and Sons own thirty-five vessels, including twenty-four beam trawlers, almost a twelfth of the entire global fleet of these destructive beauties. Apart from rising fuel costs, there are no restrictions in place on the size of the weights they use. To cause the utmost damage, 10-tonne weights are recommended. Yank them like wrecking balls over fragile deep-water targets. British waters hide some of the best coral formations in the world, comparable in fact to their more celebrated counterparts in the southern hemisphere whose plight has triggered international attention. Only discovered in 1988, they may have taken eight millennia to evolve, but it should require less than four decades to abuse them beyond recognition. Suffice to say, get trawling before word gets out.

The Stevensons possess the tools to do just this, and although there is no evidence to suggest they will deliberately desecrate the seas, they do have form. Not only have they admitted to a £140,000 quota scam, but one of their beam-trawler owners, William Stevenson, was previously fined for fiddling his log books to conceal the area where he caught seabed-dwelling sole. It is worth noting that the company hires out trawlers, a method which might be considered if the Stevensons refuse to play ball. In the event that you are forced to hire, make sure to accidentally drop old netting over the side; even discarded at the bottom of the sea, it will snare marine life. Called 'ghost netting', it may take out only two, three or seventeen fish. But in the context of rapidly dwindling stocks, every dead fish can be viewed as a minor triumph.

Net gain

Naturally, fishing can be utterly wasteful. Trawlermen have been caught hurling up to half a catch back into the water. Too small. Too oily. Too wishy-washy. Too fishy. Supermarkets, where 90 per cent of the fish will end up, like their fish a certain way. So do you. Dead. The European Commission estimates that 'discard' accounts for seven in ten deaths of fish in some waters. Scientific evidence proves that the waters around us need protecting from so-called damaging fishing practices, but these warnings continue to be ignored. 99 per cent of British waters have not been afforded any protection whatsoever. And while 99 could be 100,

BOTTOM TRAWLING

spirits can be raised with a quick international comparison of marine policy. Latvia, with barely 300 miles of coast, has more marine reservations than Britain, whose coastline is twenty-six times the size. Australia has declared one-third of the Great Barrier Reef a 'no-take zone'. New Zealand has twenty-eight protected sea zones. The UK? Just three. A hat-trick of tiny reserve zones, introduced in the last quarter of a century. And one, Strangford Lough, has been almost totally trashed by scallop dredgers.

Sea change

It could have been very different. Four years ago the eminent Royal Commission on Environmental Pollution declared that 30 per cent of UK waters were to be made reserves. For the first time, rules were put in place to prevent exploitation of the sea by fishermen, oilmen, dredgers and energy farmers; a Magna Carta for the beleaguered British fish. Ministers described the protection of the seas as 'one of the biggest environmental challenges' facing the world.

But in politics, as at sea, the weather changes quickly. Suddenly, your plot to despoil the nearby seas was back on. The Marine Bill, promised in the government's manifesto, sank without trace in late 2007. Attempts at saving the seabed stand no chance against the influential fishing lobby, whose persistent success in shaping the debate ensures there is little need for you to wade in. They argue that, if fishing bans were put in place, livelihoods would be lost. A valued and honest tradition would be gone for ever. Care is taken not to dwell on the 40 per cent of commercial fish stock that have fallen below sustainable levels in the North Sea and evidence that only 16 per cent of north-east Atlantic fish stock are within safe biological limits. If this topic does crop up, the fishermen attack oil platforms for disturbing the seabed. This is always their trump card, despite no recorded instance of a platform being dragged miles across the seabed every week.

And so, the unseen destruction continues to be ignored, even sanctioned. But there is no room for complacency. Under no circumstance should any footage of the current state of the seabed get out. A single image of the not-yet-celebrated British coral reefs could single-handedly undo all this hard work, particularly if it showed any hint of damage. In Norway, a snippet of seabed destruction so horrified viewers that the government went soft and completely banned trawlers from coral-bed areas.

Groovy, baby

Despite its success rate, trawling is bloody hard work and other means of wanton desecration are worth exploring. Construction, particularly, should not be forgotten. Gratitude should be extended to the Queen for giving up some of her much-prized estate, not only to facilitate climate change, but also to ensure undocumented devastation of the seabed. It all began with forecasts that Britain's population growth meant acute housing shortages. The government responded by ordering 200,000 new homes to be built in the south-east of England. They were to be in 'sustainable' eco-towns, an international showcase on how to build and simultaneously keep one eye on the needs of the planet. Aggregate was needed from somewhere, somewhere discreet. Quarries were too ugly, an obvious no-no. In terms of mines, they don't come more hidden than an underwater pit in the Median Deep, a crater halfway between the coast of Sussex and France which handsomely doubles as a valuable nursery for fish. From here, millions of tonnes of sand and gravel could be excavated, well away from the twitchy-curtain brigade. In 2006 more than 24 million tonnes of building material were quietly taken from the Queen's subterranean estates, a fifth of all aggregate used in Britain and a 12 per cent increase from the previous year.

Tiresome concerns persist with regard to the scooping of the so-called Queen's Bottom. Some bores liken it to ripping away a metre of topsoil from the best vineyard in Bordeaux. In a rather hysterical letter to the European Commission, the French authorities protested, claiming 'an irreversible change' to the

BOTTOM TRAWLING

Channel would ensue, leaving parts devoid of life. Typical. You would have thought that, more than anywhere else, beneath the waves you would have carte blanche to really indulge in destructive activities. Nonetheless, don't be deterred from investing in property. The house-building boom is just beginning and 325,000 acres of Britain's seabed has been quietly licensed by the Crown Estate for further dredging. Sand and gravel is used liberally to make concrete, which is responsible for 8 per cent of greenhouse-gas emissions. So press ahead with that new dining-room extension, and vow to eat only the freshest of bottom-dwelling fish at your mahogany table.

WHAT'S THE DAMAGE?

* Large pit in Channel deepens. French say this and that. No one cares. Government doubles house-building projections. **Likely.**
* A fifth of Britain's seas are designated marine reserves in response to new studies that reveal crashing fish populations. **Minute possibility.**
* Beam trawlers are phased out, but only after Stevenson & co are caught fiddling figures again. **Implausible.**
* New underwater images of Darwen Mounds reveal they are criss-crossed with fresh handsome scars. Fish sales are unaffected. **Foreseeable.**
* Just 10 per cent of fish stocks are deemed sustainable by 2013. Due to scarcity, four-fifths of species considered too unethical to sell. **Almost certain.**

Likelihood of beam trawlers being banned by 2015: 61%

Palm feeder 18

Hands up who's feeling hot

AGENDA

* Power your car with 'friendly' fuel
* Don't send Rover into a rage
* Burn peat bogs the size of Switzerland
* Have a break, have a KitKat

World leaders were in a quandary. Surely there was a way to cut greenhouse gases without antagonizing the powerful car industry. They didn't want Fiat having a fit or Jaguar making a run for it. Attempts began to replace petrol with a cleaner option. The breakthrough came with news that palm oil, a cheap and plentiful biofuel, could be burnt in engines without increasing carbon-dioxide levels. People could keep their cars, and governments could meet climate-change targets. Biofuel was hailed as the new miracle cure.

When the EU announced that, by 2010, 5 per cent of all transport fuel sold should be biofuel, astonishing results were promised in the fight against global warming. A British government agency, set up to promote biofuels, squealed that its use as a petrol alternative would save 3 million tonnes of carbon dioxide a year. It looked as if the future truly would be a putrid shade of green. As it turns out, the claims made for biofuel were untrue; it wasn't even close to being a panacea against planetary decline. In fact, the use of palm oil actually increases climate change, and not just marginally – several hundred times more than petrol does. But with a flamboyance bordering on genius, world leaders continued to promote the world's most carbon-intensive fuel. Way to go.

For peat's sake

Like a giant underground sponge, the peat swamps of the Indonesian rainforest have soaked up massive amounts of carbon to become one of earth's richest carbon stores. In normal rainforest, more carbon is stored in soil microbes than in the leaves and branches of trees. In peat wetlands this is magnified many times, with soils extending metres deep. When burnt, these swamps release vast amounts of carbon dioxide. Peat lands in the Indonesian province of Riau, an area the size of Switzerland, hold 15 billion tonnes of carbon – equivalent to global greenhouse-gas emissions for the entire planet for an entire year. A peach of a target if ever there was one. Torching this globally vital bog is a desirable part of any strategy to violate ecosystems. But sit back and put away your matches. No action on your part is needed. The great peat swamps of Indonesia are already being burnt at the behest of the British government and fellow European leaders.

Almost as soon as the EU Biofuel Directive was passed, Indonesia declared its intention to supply Europe with masses of palm oil. Europe, delighted that a ready supply of 'environmentally friendly' fuel would be coming its way, gave the thumbs-up.

Deforestation commenced overnight. 40 million acres of rainforest were earmarked to propel Europe's 270 million cars. There began in Asia one of the greatest releases of carbon in recent history.

The intense burning and draining of the peat bogs has seen little old Indonesia slither up the evil-green-guy-gauge like greased lightning, from a barely-in-the-game twenty-six to an enviable number three. Only China and the good ol' US of A produce more carbon dioxide. In 2007, the peat bogs released a commendable 1.8 billion tonnes, eight times more than the entire UK power industry managed to pump out. And not satisfied only with bronze in the CO_2 charts, Indonesia has also been recognized in the 2008 *Guinness Book of Records* for its Herculean efforts to win the coveted 'fastest rate of deforestation' title. Beat that! To stand any chance of winning, a country really has to go for it. The archipelago managed to clear a meritorious three hundred football pitches of bog an hour. The peat lands might cover less than 0.1 per cent of the earth's surface, but already they are responsible for 4 per cent of global emissions. At present, Indonesia has around 6.5 million hectares of palm-oil plantations and hopes to double this during the next five years. It is an enticing prospect, and everything must be done to ensure that it is not derailed. Luckily, as long as the British government continues to push biofuels as the ultimate eco-friendly option, there is no chance of that.

Fan the flames

As concerns crept in, Dutch scientists decided to determine the ethicality of palm oil. They were dumbstruck by their results. Biodiesel from palm oil was not just 'a bit' worse than ordinary diesel at increasing climate change. Not even twice as bad. They found that Europe was demanding an eco-fuel that caused up to ten times as much climate change. Every tonne of palm oil results in up to 33 tonnes of carbon-dioxide emissions, compared to the three that petroleum produces. And evidence kept piling up, causing some to question the government's laudable, unswerving

PALM FEEDER

commitment to biofuels. Work by esteemed Nobel Laureate Paul Crutzen showed that the official estimates ignored the contribution of nitrogen fertilizers, which generate considerable amounts of greenhouse gases almost three hundred times more harmful than carbon dioxide.

Parliament's committee for the topic had seen enough. In January 2008, MPs demanded a biofuels moratorium. But this went nowhere and a fortnight later the European Commission, backed by a consortium of motor interests, including BP, reinforced its support for biofuels by backing a long-term development plan. Days later such dedication received another challenge. An international study found that converting ecosystems for biofuels could release hundreds of times more carbon than from fossil fuels. Indonesian rainforests received a special mention; their destruction released so much carbon it would take 423 years to pay off. Its environmental con-trick fully exposed, the government kept its head down and held its nerve. Of course, they may have known all along that this might happen, that eventually the truth would out. It was realized long ago that using palm oil to tackle global warming was like pouring petrol on a chip-pan fire. Bicycles not biofuels, advisers had urged. As early as 2005, the British government had identified the 'main environmental risks' as hailing from the large palm-oil plantations in Asia. This in the very same report which agreed to the European Union's stricture that a decent proportion of fuel sold should come from plants.

The automobile industry had won, once again emerging as an indefatigable accomplice to all those harbouring hopes of runaway environmental breakdown. Without its behind-the-scenes influence, Indonesia's famous swamps may never have been turned into a gigantic vegetable oil field. Fierce lobbying against Europe's targets for cutting carbon-dioxide emissions was led by German car-makers Mercedes and BMW. German chancellor Merkel was sympathetic, saying she would resist attempts to impose caps on emissions on all new cars. It came down to a straight choice between forcing the car industry to introduce onerous fuel-efficiency measures or promoting biofuels, which had the added attraction of reducing carbon without needing new vehicle taxes. It was a no-brainer. The number-one rule in the guidebook of political survival is to avoid upsetting motorists.

In the palm of your hand

As always, there is a way for the average Joe to contribute to the race, even if you don't drive and will never have the chance to feed your four-wheel-drive a hearty diet of Indonesian palm oil. All you need do is have a break, have a KitKat. Raise two chocolate fingers to the planet. Nestlé's premier chocolate bar – two million are sold every week in the UK – has been linked to the destruction of Indonesia's forests and peat lands. An investigation last year found that KitKats have been produced using palm oil from uncertified sources. If the prospect of chomping on Britain's favourite chocolate bar and simultaneously helping to destabilize earth is a task too far, swallow something from Anglo-Dutch food giant Unilever, one of the largest corporate food monoliths in the world. Unilever uses palm oil to bulk up hundreds of supermarket products. Nestlé claims to use 'responsible' suppliers, while Unilever maintains that its palm-oil plants derive from environmentally sustainable sources, assurances that at least the British government is reticent to give. If anything, it has led to the odd position of having to condemn the ethics of an incredible corporate hulk. Ironically, Unilever has repeatedly warned of the 'unintended consequences' of the government's position on palm oils. Helping form the Round Table on Sustainable Palm Oil, Unilever's greatest crime for your purposes might be its honesty. Inside an official report from a round-table meeting, members conceded that talk of sustainability was 'a pipe dream'.

One day the palm-oil run may end. Perhaps Europe's leaders will get bored with razing Riau to the ground. When this happens, it'll probably be too late. No less than 10 million of Indonesia's

PALM FEEDER

55 million acres of peat land have already been deforested and drained. The rest will soon follow. UN figures predict that palm-oil production will double from 20.2 million tonnes a year to 40 million tonnes in 2030. By then, Indonesia's bogs will have long dried out. In Europe, millions of eco-aware drivers will chug along in a haze of biofuel. They will feel good because they are showing that they care about their planet. Apocalypse now. As any harbinger of ecological collapse will tell you; once the do-gooders start killing the planet, the war is almost won.

WHAT'S THE DAMAGE?

* European governments confront scientific reality and ban unsustainable use of palm oil. **Meagre possibility.**
* Indonesia, realizing that it is destroying one of its most precious resources, calls a halt to biofuel supply. **Improbable.**
* Europe, panicking it will never reach climate-change targets, increases percentage of petrol that derives from biofuels and palm oil. **Likely.**
* Torching of rainforest is quicker than predicted. Huge fires started by renegade companies consume much of Indonesian rainforest by 2015. **Plausible.**
* Automobile lobby and supermarkets unite to woo environmentally sensitive customers. They neglect to mention that this 'green fuel' is palm oil and accelerates climate change. **Certain.**

Likelihood of Indonesia's rainforest largely disappearing by 2020: 71%

Eau naturel 19

Don't slake your future on it

AGENDA

* Boycott the tap
* Hydrate on the hoof
* Use oil like water

Were it not true, it would stretch the comprehension of even the most visionary planetary destroyer. The world's most widely available resource is delivered straight into your home, but you choose to ignore it, opting instead for an inferior product imported from up to 12,000 miles away, at an infinitely higher price. Genius. Bottled water: the most glorious marketing miracle of modern civilization. Like sending coal to Newcastle, only more damaging.

A strong current

Until supermarkets start selling bags of oxygen, the story behind bottled water may never be bettered. It's the most sublime example of capitalism ever seen, a multi-billion-pound market created where none could ever have hoped to exist. And the environment is paying the price. Thirteen billion plastic bottles were bought in Britain last year. Less than 2.7 billion of these were recycled, the rest incinerated, dropped in fields or gutters, or dumped in landfill sites where they will take half a millennium to biodegrade.

An estimated 1.5 billion barrels of oil a year are required to produce all these plastic bottles, enough to power 100,000 cars for a year. Grab that litre bottle in front of you and take a proper look. If you factor in energy costs of production, transport, refrigeration and disposal, you may as well fill the bottle a quarter full with oil. The bottled-water industry emits more than 2.5 million tonnes of carbon dioxide a year. Some may label bottled water ostentatiously useless, but for your designs its popularity is a work of art.

Thirst for destruction

A blue mountain and a flawless white cloud. Evian. Its message on a bottle. So pure, so tranquil, so splendidly healthy. A beautiful model sashays across your television carrying a bottle of water. She is walking and carrying water at the same time. Wishing that you, too, could hydrate on the move, you start buying bottled water. Immediately, you are impressed by how easy it is to carry. No more lugging along the kitchen sink whenever you fancy a drink. Word of this ingenuity has spread across the world, and bottled water has become the world's fastest growing drinks sector, with hefty rises expected over the next five years. Sales have reached more than £2 billion a year in the UK alone. According to government figures, last year Britons drank 965 million litres. You can be confident that the environment, like the forgotten tap water, will soon be going down the drain.

So how have millions of people been convinced to pay over the odds for something they already have, quite literally, on tap? A slick campaign is essential. In the summer of 2007 the bottled-water big guns – Coca Cola, Danone (which owns Evian and Volvic) and the British Soft Drinks Association industry – decided to redouble their marketing nous with a new weapon: the Bottled Water Information Office. Its website displays a bottle of super-see-through liquid sat next to a shiny apple and a swish laptop. A mission statement reads, 'Bottled waters offer the best choice of all for those looking to quench their thirst and rehydrate with the ultimate in healthy convenience.' Although perfectly accurate, you must hope that in these on-the-go days no one has the time to analyse such wording or to remember that the stuff from a tap has similar qualities and is equally capable of slaking a thirst. As for convenience, if used correctly, a tap can also be used to fill a bottle.

Marketing mania

The Bottled Water Information Office reveals that bottled water conforms to the 'very highest standards of hygiene, provenance and sustainability'. Sustainability is a tricky claim to support, and relies on the stupidity of millions, who fail to realize that tap water might be more environmentally friendly than bottles dragged halfway across the planet. Although the PR strategy of the BWIO may have taken it a little far this time, sales will undoubtedly continue to rise. The BWIO may sound benign, but their attack on the tap is commendably vicious. They recently saw fit to promote the vague findings of a report, claiming that humans had spread chemical contaminants and that 'traces had even been found in tap water'. Soon enough they issued the warning that those who refused to acknowledge their product risked serious health problems, including 'poorly conditioned hair and skin'. Millions, it advised, were putting their health at risk by not drinking bottled water. In the wet, relentlessly dismal summer of 2007, the industry warned that heatstroke was a risk and generously provided advice to keep us alive and drinking: 'Keep a bottle of water with you at all times. Don't wait until you're thirsty to drink water – thirst is a sign that you are already dehydrated.' Finally, for good measure, they added that road rage was linked to not drinking sufficient fluids.

With the world sufficiently informed of its merits, sales of bottled water are certain to sustain themselves and, at this juncture, there is no reason for you to meddle with the clever marketing strategies that continue to hoodwink millions. Among them are the denizens of the House of Commons, who have vowed to keep using bottled water on their premises because of its 'cost-effectiveness'. MPs and staff quaff 250,000 bespoke bottles bearing the portcullis gates of power, each costing £1 a litre, and which they rarely indulge to finish. When asked if consumption of bottled water was setting the right example to the public, one MP for the House of Commons Commission, which investigates issues inside parliament, fittingly explained that using taps was just not viable. Perhaps if they want to set an inspiring example, they should consider switching brands. Word is that Waiwera, a delightful tipple undeniably worth the 12,000-mile journey from its New Zealand spring (while Evian is ferried a relatively puny 460 miles), is the next big name on the bottled-water scene. There is no excuse not to buy Waiwera. In a blind taste test of twenty-four different waters, senior sommeliers judged it to possess a smashing taste and, at £9 a litre from Claridges in London, not unreasonable value, especially when you factor in journey length, commensurate transport costs and carbon-dioxide emissions. It is hard to believe the BWIO has yet to put its marketing muscle behind this brand. In the same taste test, tap water, annoyingly, came third. Fortunately, it lacks the sophistication and hype to compete with its ecologically destructive bottled rivals.

Bottle it

If all that doesn't quench your thirst for environmental destruction, another option is to create your own bottled-water brand. Buy some land above an aquifer and apply to the Environment Agency for a licence to tap it. Build an inefficient factory to bottle the stuff, add some trace elements that won't make a jot of difference to human health, and then flog it, preferably to somewhere like Ulan Bator, the capital of Mongolia. Plaster the bottle with an image of an upland valley wreathed in glittering frost, call it Vivacity and you're off. Even better, just take liquid from the pipes beneath Kent, throw in a sprinkling of cancer-causing chemicals, give it an odd name and stick a £7 million marketing campaign behind it in the assumption that millions will relish its convenience.

That's what Coca Cola did. The company took Thames Water from the tap in their Sidcup factory, Kent. Then they put it through a 'highly sophisticated purification process' based on NASA spacecraft technology but uncannily similar to that used in home water-purification units. They added calcium chloride for an 'elegant taste profile' and pumped ozone through it, a masterstroke that changed the harmless compound bromide into the cancer-causing chemical bromate. Satisfied with its product, Coca Cola decided a mark-up from 0.03 to 95 pence per half-litre was reasonable. They called it Dasani and had the nerve to describe it as 'pure'. Swish-looking bottles of Dasani, containing up to twice the legal limit for bromate, were distributed to shops. Within months the product was withdrawn. Since then, Coca Cola seems to have changed tack, even warning that it will attempt to recover and recycle billions of the plastic bottles it uses. If successful, a global boycott of Coca Cola products might need to be considered.

Elsewhere, a vast proportion of bottles are still made from a plastic called polyethylene terphthalate, which contains traces of toxins called antimony. This plastic is used because its bright, transparent sheen complements the virtuous contents inside. The toxins leach from bottles into the water in the same way that water absorbs flavour from a teabag, but safety scares have yet to deter the public. So has the fact that 99.96 per cent of UK tap water meets stringent standards and that the 0.04 per cent that fails is still safe to drink.

Climate change will ensure that bottled water remains the fastest growing sector in the global drinks market. Investors could do a lot worse than target a sector which is set to expand in a hotter world. The stifling European summer of 2003, which killed more

EAU NATUREL

than 20,000, may well become the norm and the BWIO's well-meaning health messages will only become more pertinent, its lobbyists ready to capitalize whenever anyone dies of heatstroke.

Almost two billion people on earth have no access to clean water or sanitation. Some argue that the UN's goal, to halve the number of people without access to clean water by 2015, could be achieved with less than a third of the annual amount spent on bottled water. They are missing the point. In a world obsessed with image, bottled water will become the ultimate lifestyle accessory. It makes no sense and is dreadful for the planet, but when it boils down to the battle between the individual and the environment, there will only ever be one winner.

WHAT'S THE DAMAGE?

* Bottled water becomes the new pariah. Widespread boycott leads to sales of reusable cylinders marked, 'I'm not a plastic bottle'. **Possible.**
* Heatwave grips Europe in 2011, with 22,000 victims recorded in July alone. Bottled-water sales treble. **Plausible.**
* A series of safety scares involving bottled water fail to dent sales. **Predicted.**
* Current stipulation that a fifth of plastic bottles should be recycled expires. More stringent regulations are not forthcoming, due to 'practicalities'. **Probable.**
* Bottled Water Information Office reveals that its product makes men better lovers. **Likely.**

Likelihood of bottled-water sales doubling by 2015: 77%

Not 20 soya good

A tree-free tomorrow

AGENDA

* Support soy production
* Call in the cattle
* Raze the rainforest
* Watch the planet breathe its last

Amid the daily smorgasbord of celebrity shenanigans, US elections and ministerial hiccups, a small news bulletin slipped out almost unnoticed late on 23rd January 2008. Brazilian government scientists had just received a remarkable set of images revealing that the destruction of the Amazon had inexplicably surged. More than 300,000 acres had somehow disappeared in the preceding four months. Poof. Gone.

Even in the Seventies, no one had ever recorded such a rampant rate of deforestation. The trees of the Amazon produce much of the world's oxygen and diligently hoover up millions of tonnes of greenhouse gases from the atmosphere. This forest, which sprawls over 5 per cent of the planet, is critical to the planet's survival. From the fuzzy images, it became evident that the most pronounced damage had occurred in the massive Brazilian frontier state of Mato Grosso. There, obscured beneath tremendous smoke plumes, huge tracts of land smouldered ready for soy plantations. The lungs of the world weren't looking particularly healthy.

For a while soya had seemed the answer to everyone's prayers. Massive plantations chomped lustily into the heart of the world's largest remaining rainforest. Europe had begun ordering two-thirds of Brazil's rainforest-ravenous soya to feed to livestock destined to become cheap meat in supermarkets and fast-food outlets. Every part of the British food chain contains traces of the stuff. A fifth of the Amazon rainforest has been destroyed in order to grow it, an area the size of five football pitches lost every minute over the last decade. Admirable work indeed, but a new dawn of destruction may be upon us.

Maggi, Maggi, Maggi

In the city centre of Rondonopolis, Mato Grosso, stands the imposing white-washed headquarters of the Amaggi Group. These are the plush offices of the man who will help to deface the Amazon. Blairo Maggi's credentials as a purveyor of planetary ruin are beyond reproach. No one can preside over the fastest rate of Amazonian deforestation without receiving some plaudits. Maggi is the world's *Rei da Soja*, the King of Soy; his offices are the nerve centre of the planet's largest soya-producing organization. Like all conscientious businessmen, the 48-year-old wants to keep expanding his business. And nothing can stop him. This millionaire father-of-three is also the governor of Mato Grosso. He has revealed plans to develop 40 per cent of his state for agriculture, doubling production in a single year.

Maggi is not shy about his exploits. When questioned about the destruction caused by his soy empire, he could have said: sorry for attacking the richest ecosystem on earth; I was only trying to fulfil the economic dreams of my beloved Brazil. But he answered like a man, saying boldly, 'I feel not the slightest guilt.' His soy kingdom has grossed more than £300 million per annum in recent years. Facilitating his success is an impeccable range of political contacts. The state's environment minister has been arrested for taking kickbacks. Dozens of Mato Grosso's government officials, paid to protect trees, have had their collar felt for accepting bribes to expedite unlawful rainforest destruction.

Tout, tout, tout

The business proposal to put to Maggi is straightforward. Foreign investment from Europe will fund a fresh generation of planters, processing units and lorries to hasten soya production. Land will be bought from Maggi at a flat rate in exchange for generous returns to his company. British and European agribusiness will be politely courted to profiteer from the soy boom. The first aim is to convert Mato Grosso, which, with delicious irony, means 'dense forest', into a whopping soy field the size of Europe. Maggi has also groomed excellent soy contacts in neighbouring state Para. Plantation owners will be ordered to take down every last tree and shrub as part of their contract. Once Para has succumbed to soy, 60 per cent of the world's most celebrated ecosystem will have vanished. From there, march westwards into the rainforest's untouched depths. By 2015 those mighty lungs will be little more than wheezing gasbags. Admittedly, the destruction of the Amazon is audacious, but with Maggi on your side, it's eminently doable. International opprobrium will be inevitable. But Europe and the US are hooked on cheap meat. And what the consumer wants, you must make damn sure they get.

Breathe your last

The Amazon's global status is a magnet for carping greenies and postponing the international outcry for as long as possible is key.

Don't forget the Amazon's traditional means of silencing those who stand in the way. Mato Grosso and Para are violent states, and refuge to scores of *pistoleiros* (hitmen) hired to protect the soy plantations. Forget São Paulo or Rio; the murder capital of Brazil is a charming little Mato Grosso town called Colnizia. Police in the region have said that a 'hit' here can be bought for £20. In Para, mercenaries can be sought from the nearby town of São Felix do Xing, where motorcycle helmets are banned due to the frequency of contract gunmen. Crimes are hardly ever investigated, and the murder of environmentalists in the Amazon is long established. American nun Dorothy Stang was famously killed in Para after years of campaigning against rainforest destruction. She was shot point-blank in the face by two hitmen hired by a nearby rancher for almost £10,000, then left face-down in mud, Bible by her side. Hours later, a local environmental activist was killed by gunmen in front of his wife and children, and ten days on another Brazilian environmentalist was shot in the head after speaking out in defence of a Brazilian nature reserve.

There is no room for sentimentality in the great new soya push. A provisional list of targets has been prepared for you and, one by one, the critics of the soy expansion will be quietened by the laws of the jungle. Edilberto Sena might be among the first. Already, this prominent Catholic priest has received death threats for publicizing the illegal activities of soy farmers in the Amazon. He would be advised to heed the warnings. Renowned Dutch primatologist Marc van Roosmalen, named by *Time* magazine as one of their 'heroes of the planet' for his pioneering studies in the Amazon, also needs to learn to keep his mouth shut. The 60-year-old rightly fears for his life after unidentified gunmen threatened him for spilling the beans on certain soy companies. Rumour indicates he sleeps in a different place every night to avoid being killed. He should know by now that no one escapes in the Amazon.

What a wheeze

It's important to make friends in high places. The very people tasked with protecting the Amazon might be persuaded to come on board. Almost fifty officials belonging to the federal environment protection agency in Mato Grosso have been arrested for accepting kickbacks, and there are surely more corrupt officials where they came from. Mato Grosso's most senior official, Hugo Jose Scheuer Werle, the leader of Brazil's premier environment agency, was arrested for allegedly accepting £100,000 from illegal loggers for providing them with falsified documents. He was not convicted. Even more promising is news that the state's very own environment minister, Moacir Pires, was accused of aiding a logging ring that cleared 52,000 football pitches of rainforest. He was also arrested, but like Werle, not convicted.

With the support of such friends, the conversion of Mato Grosso and Para to a soya desert will be a cinch. By then, Maggi will have capitalized on government backing and soya-export companies to complete the 1,000-mile paved BR163 – the Soy Highway – running from Mato Grosso, through Para and on to the gigantic Amazonian port of Santarem. It is being illegally constructed by a US soya giant. There are few better ways of hastening rainforest destruction than a good road, particularly one which will create a 25-million-acre swathe through the region. And, in case you're wondering how it will affect the Amazon's protected forest reserves, the road's bewildering scale renders policing practically impossible. The country's environmental protection agency will have as little as fifty officials looking after an area the size of France.

Bank on Maggi

Time, now, to extend some appreciation to the British taxpayer. Back in 2002, as Maggi was probably wondering how he might ever fulfil his dream of Amazonian destruction, the International Finance Corporation – the private-lending arm of the World Bank – kindly gave him a £15 million loan. Loaded with cash to burn,

Maggi immediately financed 900 soya planters. The IFC classified the loan as category B, meaning that any environmental impact which might arise could 'be avoided or mitigated'. Britain recently overtook the US as the largest donor to the World Bank, promising more than £2 billion from July 2008 to 2011. Shortly beforehand, it emerged that the World Bank was a principal backer behind an explosion of cattle-ranching in the Amazon, second only to soya in terms of effective ecological destruction.

Brazil has become the leading beef exporter, outstripping all twenty-five EU members put together. Around 74 million cattle live in the Amazon basin, four for every other species found there. Even so, the IFC agreed more cows should be squeezed in and gave one beef processor £4.5 million to upgrade its slaughterhouse. An internal IFC study confirmed that expansion of a single slaughterhouse could result in a loss of 750,000 acres of rainforest.

Events even further afield may also hasten the demise of the Amazon. The Amazon's impressive transformation to a soya plantation only really took off after the first case of Bovine Spongiform Encephalopathy was identified. Brazil continues to supply an untainted source of cattle feed and, if another BSE crisis were to strike Europe, Maggi and his mates in the Amazon would clean up. Soya prices would rise overnight: it would be curtains for Mato Grosso and Para more quickly than earlier predictions. Another CJD death linked to European beef would also do the trick.

Already, Britain is the sixth largest importer of Brazilian beef, procuring more than 80,000 tonnes in 2007. The commensurate deforestation means that Brazil has gleefully soared into the top four of the planet's best carbon polluters. Admittedly a tough ask, with China, US and Indonesia all performing stirringly, but if BSE came back into town then the Amazon might yet be a chart-topper.

So, the trees should feel duly worried. The only dot on the horizon is that Maggi has entered into talks to sign a two-year moratorium on further soya production. He is a smart operator and you must feel confidence that his motives are driven by naked tokenism. Maggi must sleep well at night knowing that he is loved by so many. All along the BR163, his chubby face smiles from billboards, and he has been widely tipped to make a bid for the presidency of Brazil. Why not? Is it a crime to give people jobs whilst feeding the world? Keep eating meat, it's the least you can do to show your appreciation.

WHAT'S THE DAMAGE?

* Western countries set up global fund to preserve the Amazon. Its ambition is comparable to the Marshall Plan. **Not a chance.**
* Maggi, already a millionaire several times over, halts expansion of soya production in Mato Grosso, citing environmental reasons. **Forget it.**
* World Bank refuses to fund anything harmful to the Amazon, citing ethical reasons. **Implausible.**
* Europeans continue consuming cheap meat, oblivious to or unbothered by its role in puncturing the so-called lungs of the world. **Very likely.**
* The Brazilian government orders a series of huge police raids in Mato Grosso and Para. Hitmen and illegal soy planters are imprisoned. **Unlikely.**

Likelihood of Mato Grosso and Para being destroyed by 2015: 83%

NOT SOYA GOOD

21 Sea of change

Urea the new panacea

....

AGENDA

* Enjoy some algae entertainment
* Choke the oceans
* Suffocate marine life
* Know where your waste is going

Waste not, want not

It was some field, flat and featureless as far as the eye could see. Inside the cockpit of the Qantas Boeing 747, the pilot glanced at the co-ordinates. Scratching his head, he double-checked the flight path. The plane was 10,000 metres above the North Pacific, yet the ocean had mutated into land, a muddy-green agricultural prairie.

Later, the pilot would hear that a gigantic mass of plankton and algae had blossomed in the North Pacific. Conceived by some as a desperate method of tackling climate change, it had burgeoned out of control. Despite this, the deliberate growth of algae and phytoplankton, the microscopic plants that form the lowliest rung of the marine food chain, continues to be advanced as a way of preserving the planet. The perceived wisdom is that the algae absorbs carbon dioxide then dies and sinks to the seabed taking with it the gases it has absorbed. Humans can pump extra nutrients into the ocean, to create more algae and thus extract more carbon dioxide from the atmosphere. Simple.

Some argue that this technique will starve the seas of oxygen; others warn that the reduction of carbon dioxide is open to question. The International Panel on Climate Change describes ocean fertilization as 'speculative, unproven and with the risk of unknown side-effects'. And what magnificent side-effects. Studies reveal that these ocean prairies could even exacerbate climate change. Yet another glorious way to f**k the planet. Encourage this large-scale experiment and you will ensure that the oceans are sucked dry of oxygen. From above, they may resemble a lush land of plenty, but the abyss below will resemble a watery graveyard.

Urea – eureka!

Day after day, the algae kept growing until it covered an area the size of England. Panicking, those monitoring the zone ordered a submarine survey. Its findings confirmed their worst fears. Apart from several crabs and marine worms scrabbling about on the ocean floor the sea contained nothing. Where were the fish? The whales? Above, weakened sea birds searched vainly for food that no longer existed. The 2006 dead zone off the coast of California lasted for nearly seventeen weeks, longer than scientists had ever predicted. Their quest for an explanation led them to the west-coast current, which had carried an up-swell of waters unusually rich in nutrients. Phytoplankton levels had boomed accordingly. The aquamarine sea mutated into a stodgy green-brown soup. The algae then died. Oxygen levels plummeted to almost zero.

When news of this dead zone first reached Europe, it was hailed by the writers of Armageddon handbooks as a fabulous break-

SEA OF CHANGE

through for those plotting the Anthropocene, the destructive era of man. In the months that followed scientists started to plan a series of large-scale trials. Small doses of iron were found to encourage the algae. But this compound already existed in 80 per cent of the ocean and, sadly, adding fresh quantities in fertilizer form would not give a sufficiently damaging return. Another method was needed.

The latest proposal relies on pumping urea into the sea. Millions of gallons of this foul-smelling by-product are excreted every day by the human race. By practically pissing into the ocean you can really screw things up for Mother Nature. Urea has been coveted by farmers for decades as a nutrient-rich fertilizer that makes crops grow. When dumped into the ocean, it has pretty much the same effect, turning oceans into thick, choking blankets and compromising any hope marine life ever had of breathing.

Let me bend urea

The evocatively titled Ocean Nourishment Corporation of Sydney is proposing to dump 500 tonnes of industrially produced urea into the sea between the Philippines and Borneo in order to grow algae, with the promise of absorbing carbon dioxide. There is a lot to learn from this bunch. Their chief executive, Ian Jones, has been quoted as saying that, in the same way that humanity has cultivated land, their plans are 'like practising agriculture at sea'. The ONC's plans were unveiled during a UN treaty meeting in London just over a year after the dead zone off California came to attention. The meeting fundamentally approved the plan, although a spokesman for the Department of Environment, Food and Rural Affairs later confirmed that such schemes were 'potentially high risk'. Less than a month later, the London Convention urged countries to use extreme caution when examining plans for large-scale fertilization operations. But it did not say no.

Your next stop is Planktos, a company based in San Francisco, just a short hop from the dead zone of 2006. Planktos wants to dump thousands of tonnes of iron in the Pacific off the Galapagos

Islands. A delightful proposition you'll agree. The very site where Charles Darwin made key discoveries about natural selection might actually become encircled by a mushy, lifeless, man-made soup. Tests show that Planktos is on to something. Half a tonne of iron has already been added to a planktonless area of the Pacific just off the islands where the theory of evolution was developed. Algae bloomed. The sea turned green.

Back in Sydney is ocean engineer Professor Ian Jones who loves urea and believes it can save the world. Ask him to discuss his project to build great pipelines into the oceans through which urea would be pumped to create massive algae blooms. Convince him that a marvel of engineering is required; scores of huge pipelines running into every ocean are the only way his solution will be taken seriously. Offer large-scale financial backing if he will exploit the access and credibility that accompany his academic eminence. Do not, under any circumstance, divulge that you hope it all goes disastrously wrong and backfires. Equally, do not mention that you are developing a system which overrides any attempt to switch the pipes off. Once the urea is flowing, you must ensure it continues to flow until the oceans can no longer breathe. Convince Jones that, to maximize returns, he should target areas where there is already a heavy density of phytoplankton. Tests reveal that even just a modest increase on existing levels can offer an excellent return.

In your own backyard

If Jones proves none too conciliatory, then exploit the thousands of sewage outflow pipes around the world that still pump waste directly into the open sea. London-based suppliers sell bags of urea crystals at £150 a tonne, sourced from China, India and Indonesia and with a minimum nitrogen content of 46 per cent. If things get desperate, dump granules directly down toilets then stand back and watch the waters bloom. Sewage leaks are also helpful, capable of releasing millions of tonnes of urea.

Re-mortgage your house (the property ladder won't take you

SEA OF CHANGE

anywhere anyway, since you're busy pissing on the future) and spend £150,000 on over 1,000 tonnes of urea crystals. Previous tests reveal that a tonne of urea will affect a few hundred square metres of ocean. The pile in your garden is large enough to choke an area almost the size of Italy. Of course, everyone can do their bit. Douse gardens liberally with nitrogen-rich fertilizers, wait for the rains and hope the run-off will liberate growth-inducing compounds into nearby rivers and, ultimately, the ocean. The amount of nitrogen contained within fertilizer is stated on the packaging. Aim high. Be inspired. Look at the correlation between increased use of nitrogen-rich fertilizers and the steady, sublime growth in the number of dead zones. The United Nations records a doubling of zones every decade since the 1960s. Around 200 have now been identified, compared with 150 just two years ago. UN officials are not impressed. Algae blooms, they warn, are likely to become the principal destructive factor for the ocean, replacing over-fishing.

Welcome to the dead zone

The Gulf of Mexico annual dead zone some years covers more than 7,000 square miles and is mainly caused by an impressive 1.7 million tonnes of nitrogen fertilizer dispensed into the Mississippi river. The biggest, though, is in the Baltic, where sewage and nitrogen fallout from the burning of coal and gas combine delectably to over-enrich the sea. Dead zones frequently cluster around the coasts of Europe, South America and Asia. Satellite images have caught phytoplankton blooms drifting along the west coast of Ireland which are bigger than the island itself.

We return full circle to the science. For such ocean nourishment to be effective in tackling climate change, substantial amounts of carbon dioxide must fall to the ocean floor and be incorporated in deep water sediments. Yet – and here is the vignette supreme – experiments to date show that the actual amount that reaches the ocean bed is tiny. Other studies found that these artificially induced blooms could increase the production of nitrous oxide and methane, gases which are much more effective at trapping heat than our loyal, omnipresent pal carbon dioxide. Phytoplankton itself might have the added spin-off of heating the seas by absorbing heat from the sun that would otherwise be bounced back into space. If you require more reasons why urea is the new panacea, then now is the time to accept that you lack what it takes to join the team. Put the book down, nip to the lavatory. Aim carefully and dispose of urea in a suitably decadent fashion.

WHAT'S THE DAMAGE?

* Number of dead zones hits 250 in late 2011. Entire Gulf of Mexico - an area five times the size of Italy - is declared lifeless. **Very probable.**
* UN pushes for ban on sales of nutrient-rich fertilizers. Agricultural lobby reacts furiously. Farmers prevail. **Likely.**
* Ocean-nourishment testing off the Galapagos reported to be 'extremely successful'. Permission granted for industrial tests in South China Sea. **Expected.**
* Huge sewage leak reported in Brazil. Soon after, world's largest dead zone is spotted floating off South American continent. **Foreseeable.**
* Large-scale ocean-fertilization plans to combat climate change are abandoned in 2015 after court case proves that they damage marine life. Number of dead zones breaks 400 barrier. **Conceivable.**

Likelihood of large-scale urea projects being commissioned to fight climate change: 67%

SEA OF CHANGE

22 Arrested development

Home is where the heat is

AGENDA

* Pave over your garden
* Create a floodplain in your own back yard
* See down the eco-town

Once, Englishmen refered to their homes as castles. In your capable hands the humble family home will again become the scene of battle, the perfect base from which to launch an attack. The target, as always, is planet earth. You will renovate your existing home in accordance with destructive protocol before abandoning it to take up residence in one of the government's sparkly new eco-homes in a tranquil corner of the countryside. From there, you will drive the long way to work, returning each night to eat steak and admire the Cotswolds beneath a patio heater.

Paved paradise

You rather admire your trimmed front lawn fringed by its neatly clipped privet. It's a fine example of middle-England austerity and offers slim pickings for any local wildlife that might be foolish enough to attempt colonizing. But you want yet another 4x4 and need somewhere to park it. There is no choice but to smother your cropped grass and prim borders in a generously deep coating of asphalt. The neighbours have already done theirs, and it looks ravishing, the twinkling bodywork of four cars on the driveway offering an aesthetic way beyond anything a few shrubs can offer. A lot of people agree. Thousands of front gardens already lie beneath barrowfuls of bitumen, laid to create parking spaces for Britain's 32 million cars. Over the last decade, Londoners alone have buried enough front gardens to cover twenty-two Hyde Parks. Great minds at the Royal Horticultural Society have contemplated issuing advice to its 350,000 members on how they can build a parking space while retaining a beautiful garden. It will prove a wasted initiative. This is not about elegant colour schemes; the concreting of your garden has a greater purpose. Try it. Hide the front lawn beneath the black stuff and just sit back and wait for the rains. They will surely come. Climate change guarantees erratic weather and a greater frequency of thunderstorms.

Almost immediately, you will notice an impressive difference. Front lawns provide a vital natural sponge for rainwater in towns and cities. Without them, water runs off driveways and increases the likelihood of downpours overwhelming antiquated sewerage systems. You fondly recall the fetid stench that hovered above the Thames in the summer of 2003 after heavy rains exposed London's dwindling cover of foliage by forcing up to 600 million litres of faeces and waste into its waters.

But unfortunately, you may have missed out on asphalt this time. In an outrageous move, the government has declared war on your coveted right to suffocate your front garden. They say they want to save water and reduce flooding. New legislation allows only gravel, porous bricks, or paving, which provide better drainage than hard surfaces.

ARRESTED DEVELOPMENT

Greenfield front line

Intrigued by the government's impossible-sounding notion of building three million new homes without contributing to climate change, you find yourself becoming gradually obsessed with one of the prime minister's flagship 'green' brainwaves. You fancy moving into one of his eco-towns, a settlement of between 5,000 and 20,000 homes intended to be carbon neutral. You have inspected the proposed towns and are thrilled with what you have seen.

Your trusted friends in the house-building industry are performing precisely as you had hoped. Leading developers are, naturally, using the eco-town template to dust off long-rejected proposals and re-submit shoddy housing schemes. At least nine of the initial proposals were previously rejected by planners. Your favourites are those targeting greenbelt sites. One in particular has caught your eye. A development is planned for near Micheldever, where Eagle Star Insurance has been trying to develop a London–Basingstoke commuter settlement for some years now. Its plans are now sprinkled with phrases like 'sustainable development' and 'carbon-neutral'. A bog-standard scheme for 12,500 homes on a pristine, greenfield site has craftily transformed itself into an 'eco-town'. You wistfully think of thousands of extra cars on the road and the winning concept of sprawling urbanization.

Yet some of the plans have provoked consternation; it actually seems as if they might do some good for the environment. They might not contribute anything whatsoever to f**king up the planet. With a sense of apprehension, you realize that you will need to contact the daunting Weston Front opposition group. In particular, its leaders, Mr and Mrs Henman. Tony and Janet are furious. An eco-town of 15,000 homes is planned for the periphery of their picturesque Oxfordshire village of Weston-on-the-Green, the 400-strong settlement where Wimbledon failure and former tennis player Tim Henman grew up. The Weston Front do not deal with fools. They will do anything to win. Several fetes are already organized to fund their militant stance. As Tony, 67, says: 'The village at the moment is great, it's a lovely little community. If this goes ahead it would be completely destroyed and stolen from us.' Tim is also horrified, but no one listens to a loser.

The proposed new town, Weston Otmoor, is planned for within a few hundred metres of the Henmans' £750,000 home. You must defeat the tennis player's parents. The family has a history of failure, so take heart. Weston Otmoor is an inspired development. For a start, the proposal includes building on part of the Wendlebury Meads, a site of special scientific interest and featuring loads of wildflowers, including green-winged orchids, devil's-bit scabious, betony, cowslip, and dyer's greenweed. Briefly, you wonder how civilization could plough on were dyer's greenweed to go. In addition, the proposals could generate an extra 8,000 car journeys a day. Severe traffic jams on already congested roads are always to be welcomed.

Weston Otmoor looks pretty damn good on paper. So what if the likes of the Wildlife Trusts have already suggested it will destroy local wildlife and the environment? It is too early to say, but you remain hopeful that you will acquire a home in eco-town heaven. You imagine windswept 'business parks' dumped in the sticks; cheap, identikit office buildings plonked down by some godforsaken motorway junction, and Legoland streets abandoned by public transport. You foresee that 'Tescotown' will eventually have two-thirds of the shops on the main street derelict in the shadow of an out-of-town supermarket. You hope that, as they normally do, ministers will have the sense to listen to developers on matters of sensitive planning issues, and together decide on an empty square mile of greenfield countryside. More than anything, you hope that the UK sticks to its house-building target of 240,000 homes per year. By this stage, you have already been reassured that some of the proposed eco-towns do not include homes already agreed under regional housing plans and are merely extra homes in the countryside. Some are on the edge of the Cotswolds, some

ARRESTED DEVELOPMENT

straddle so-called Areas of Outstanding Natural Beauty. Others are proposed for areas where there is already a housing surplus and thousands of empty homes lie vacant, stuck on the open market. Extol such excess; there's always a plus in surplus.

Home comforts

While you wait for your concrete-festooned delight in suburbia to be snapped up by someone who shares your appreciation of owning a lot of cars, you might want to consider modifying it in other ways. 27 per cent of Britain's carbon emissions are produced at home. Without further ado, you open all the windows. Turn up the thermostat. Destroy saucepan lids to ensure the maximum amount of energy is used when cremating stews and to create less washing up. Hastily remove any trace of insulation, which can reduce heat loss by a fifth. Get into the loft and have a look around: anything you suspect of trapping heat, rip it out. Of course, removing the roof is the preferred option, but this will only lead to late-night arguments with your partner, and even f**king up the planet is not worth that. How else would you procreate to weigh down the global population? Burn your curtains to ensure maximum heat loss at night. Form vigilante groups to sabotage any homes daft enough to have installed solar panels or windmills. Photovoltaic panels are particularly vulnerable to a stray object lobbed from a nearby vantage point.

Wasting electricity is one thing, but to be considered truly ecocidal you must squander as much water as feasibly possible. Install power showers, dishwashers, and washing machines. Develop OCD and become obsessed with washing: every time you so much as touch yourself you find yourself furiously scraping your body in the most forceful shower in the neighbourhood. Currently, you use a daily average of 150 litres per person per day, compared to the 130 litres demanded by the Germans and Dutch, who might drink less tea than you and take hygiene less seriously.

Once you've primed your property for the market, if you cannot find a buyer for your house immediately, do not panic. Occupy yourself by making sure you resist all attempts to convert your dwelling into a low-carbon living place, all the while hoping that the government does not copy Spain and begin offering subsidies for old homes to become more energy efficient. Remember: everything starts at home.

WHAT'S THE DAMAGE?

* Campaign launched in 2010 to protect the front garden. Asphalt becomes a suburban taboo. **Probable.**
* Designers launch cut-price zero-carbon house which promises to reduce electricity needs and save space. Its design flaw is that it is little larger than a wendy-house. **Maybe.**
* North-east property developer David Abrahams, investigated by the police after donating more than £660,000 to the Labour Party under other people's names, is linked to an eco-town vociferously opposed by local residents. **Plausible.**
* House-building programme increased in 2012 to cope with demand. Announcement induces fresh howls of protest over immigration levels for despoiling the green and pleasant land. **Likely.**
* Eco-towns are pronounced a success, but make no difference to the UK's carbon targets. Politicians laud them when describing how the UK leads the world in tackling climate change. **Certain.**

Likelihood of eco-homes making a tangible difference to planet by 2015: 34%

ARRESTED DEVELOPMENT

23 Green light

Trip the switch

AGENDA

* Illuminate eco-bulb hazards
* Don't shell out on new-fangled lighting
* Start a health scare
* Switch off the lights for good

You despair sometimes. Some days it feels like everyone is obsessed with protecting the planet. Green this, lentils that. As the hours wound down to Britain's first 'Energy Saving Day', you began to feel oddly restless. The organizers of E-Day had begged people to switch off electrical devices which they did not need over the 24-hour period. As expected, 27th February 2008 was one of the longest days of your life. Finally, after hours of careless energy-saving, came the moment for the National Grid to announce the results. You bit your lip. But, unbelievably, it was up. Consumption was up! Despite widespread publicity from campaign groups, electricity demand was actually higher than the 'business-as-usual' figure. Proof, surely, that you are not alone in not giving a f**k. Coming so soon after Live Earth's laughably poor audience figures, it was another development to celebrate.

But the government still seemed determined to ascertain for itself whether its public cared about all things green or whether, like them, it just talked a good game. It knew that almost a quarter of a home's electricity goes towards keeping on the lights. As a tester, it announced that it was going to ban traditional tungsten bulbs and replace them with the aesthetically challenged compact fluorescent light. The genius of Thomas Edison was recast as a crime against the climate and another means by which politicians could tell you how to run your own homes. It was a vital moment: if the British public decided to embrace the move, the first genuine steps towards a low-carbon economy would have been made. It was the thin end of the wedge. You are concerned that a comprehensive, well-planned energy-efficiency campaign might follow, a campaign that actually might make a difference. Somehow, this great light switch must be flicked off and Britain must be forced to stay in the dark ages.

And then there was light

It seemed a thankless task. You had frittered away hours trying to hatch up a plan that might stop what the *Sun* had termed 'the great light switch'. It was difficult to make the introduction of these bulbs seem anything other than reasonable. Normal light bulbs use just 5 per cent of their energy to illuminate, while eco-bulbs are sparing by comparison and last six times longer. How to stop the scheme from succeeding? The blinding flash came when you learned that these new-fangled bulbs had been known to poison family homes – the entire British population could die if they were given the go-ahead. Bingo! By some minor miracle you discovered that the great 'clean' eco-lights were actually powered by one of the most heinous toxins on the planet. Exposure to even a tiny amount of mercury could trigger severe effects on the human's central

GREEN LIGHT

nervous system. Teeth would become loosened, and hypertrophied gums bleed easily. Shucks.

Government advice was characteristically non-alarmist. Evacuate the room immediately if the bulb of a compact flourescent light should fall to the floor and break; under no circumstance use a vacuum cleaner to clear up the mess, as the machine's sucking action could spread toxic mercury droplets around the house and, better still, contaminate nearby water supplies if allowed to escape outside. This was green fascism at its finest: not only could the plan prove harmful to you and the environment, but in an era of eco-guilt, no one had a choice.

What were you to do? With exquisite timing, news broke of another health warning: doctors announced that intact bulbs aggravated a range of ailments, such as dizziness. Not to be outdone, the Migraine Action Association weighed in, alleging that eco-lights had prompted a wave of the agonizing headaches. Six million people in the UK suffer from migraines. A national scandal was in the making, and it could only get worse. Green bulbs, explained the next health warning, could induce epileptic fits in sufferers. Next up were complaints from sufferers of lupus, a chronic immune disease that prompts pain and extreme lethargy. Yet the real *coup de grâce* was still to materialize ...

The British Association of Dermatologists provided the killer blow, explaining that low-energy lighting could make people look crap. In a superficial world, little rivals the power of vanity. Environmentalists can witter on about polar bears all day, but if they learn that something might make their cheeks sag or go a bit spotty, then it's a different story. So when a leading dermatologist announced that tens of thousands of people with skin complaints might risk exposing their flaws by going near the bulbs, it was effectively curtains for the government's green dream. People were being ordered to switch to something that might save them half a penny on their bills but in return they would look rubbish. Sacrifice your chance of pulling for the needs of the planet? Not a hope in hell. Dermatologists warned that the lights could cause conditions such as eczema to flare up. Another 340,000 people with photo-sensitive skin could be affected by the first step towards an environmentally aware Britain. Rashes were a risk. Most people would rather live in an oven than risk developing a rash. And, finally – hallelujah! – there came the prognosis that, yep, you guessed it, eco-bulbs might lead to skin cancer, all caused right in your own home. And there wasn't even the flipside of a tan.

Undeterred, former mayor of London, Ken Livingstone, told everyone to calm down, just moments after warning that if they did not swap their bulbs his city's residents risked 'catastrophic climate change'. Getting into the swing of things, Livingstone then ordered a 'light-bulb amnesty': those who dumped their regular tungsten time bombs could avoid the electric chair. He glossed over the fact that some light fittings could not accept the new-fangled green bulbs. People would just have to compromise. Stairways would remain unlit, old people tumbling to their deaths in the darkened corridors of their homes.

It doesn't say anything about this on the box. You welcome the news that the Environment Agency has demanded more information be made available on the health and environmental risks posed by low-energy light bulbs. Lurid health warnings printed on packaging should do the trick, you suspect. Parents need to be told. Set up a website, campaign for the public to be informed. Okay, so the planet might be warming, but do people really know what they are risking by installing eco-bulbs in the sitting room? Write to the media asking why the EU can consider banning mercury in barometers and yet happily wrap it in a fragile glass coating and sanction its arrival in your lounge. Distribute pictures of wide-eyed toddlers gazing fondly at the poisonous illuminator. People would rather get fried by the sun than subject their little ones to hypertrophied gums.

And it's not just inside our homes that the government wants to

GREEN LIGHT

play God. It also wants to lay down its decree in the streets. Councils are starting to trial blackouts between midnight and 5 a.m. in an attempt to meet energy targets. Your cue, then, to protest at a perceived rise in muggings, robberies, dark doings. 'A perverts' paradise', you write, in one frantic missive to a local newspaper. Police can be counted on for support. Concerns over crime will always outweigh those of the environment.

There you have it: the death of energy efficiency. Fewer than three in every hundred bulbs sold is currently an eco-bulb and, as you wonder whether you actually know anyone who would risk their lives for such little gain, you wait impatiently for the next eco-bulb health warning. Impotence always works a treat.

WHAT'S THE DAMAGE?

* A small child dies after ingesting mercury from broken bulb. Complete recall ordered by safety watchdogs. **Possible.**
* A clean Amy Winehouse injured after falling down stairs. She blames 'flickering' eco-bulbs for making her lose her balance. Mass opprobrium follows. **Plausible.**
* Transformation to eco-lights happens without a hitch. **Maybe.**
* Blackout imposed by Westminster council to tackle global warming. Home secretary mugged after leaving Commons. **Likely.**
* The little-known Live Longer Bio-Medical Health Solutions consortium condemns the switch to eco-lights as ethically untenable. **Imaginable.**

Likelihood of eco-bulbs being resisted by most consumers by 2010: 78%

24
Radiating fury

Hiroshima mon amour

AGENDA

* Precipitate nuclear war
* Re-engage with the ice age
* Strike out at heatstroke
* Spread radiation throughout the nations

All this talk about boiling the planet is beginning to feel a touch unimaginative. Perhaps there is as much collateral in cooling things down a notch. As shown by the decades of concerted effort ploughed into global warming, changing the planet's temperature requires patience. Yet it is often forgotten that man has already mastered how to engineer complete climate change overnight. At the push of a button, in fact. Raise your glasses to the nuclear winter, the quickest way of all to truly f**k the planet.

The Japanese city of Hiroshima illustrated that the planet could survive the fallout of a single atomic device. But one should never underestimate man's ingenuity when it comes to engineering ways to liquidate each other, and atomic bombs have become bigger and better and more numerous than ever before.

Exterminate, exterminate

New, sophisticated climate simulations reveal that even a relatively low-key nuclear spat will trigger a chain of events from which the planet will struggle to recover. Those monitoring the studies were deeply troubled by the global impact arising from even a modest squabble between two countries that involved nukes. Scientists discovered that 5 million tonnes of soot would be spewed into the atmosphere, inducing a 1.25°C fall in the average temperature at the earth's surface. The planet's life source, the sun, would be dimmed for more than a decade. In this twilight world of ubiquitous frosts, entire harvests would fail. The planet's major breadbaskets would struggle. The *coup de grâce* involves a gaping hole in the ozone layer above Europe caused by gases emitted in the nuclear exchange. Ultraviolet rays stream through the tear in the earth's protective layer, killing crops and frazzling the seas.

Your objective is to engender the most abrupt climate change in recorded history. To achieve such a scenario, go for the burn by precipitating another Cold War. Anglo-Russian relations are so deliciously brittle it seems neglectful not to take maximum advantage. There is no better target for a full-on nuclear scrap than the vast state beyond the former Iron Curtain. Russia possesses almost 5 per cent of the world's arsenal, the majority of devices seventy times the strength of the one jettisoned by B-29 bomber *Enola Gay* above Japan that defining summer morning. Of Russia's estimated 16,000 nuclear warheads, 7,200 are fully operational. Less than a tenth of Russia's arsenal would be sufficient to trigger a nuclear winter.

The Nuclear Non-Proliferation Treaty, which once saw Russia and Britain agree to gradually disarm, has unravelled. Britain has ordered a new £25 billion nuclear-weapon programme with warheads that promise a hundred times as much deadliness than the lump that fell from the *Enola Gay*. Russia, meanwhile, is considering suspending any further reduction of its huge nuke stockpile. For the first time in decades the spectre of a nuclear winter is back. The Cold War proved itself to be a lot of hot air from frosty diplomats. This time around things will be different – the Cold War to end all wars, a conflict to end in downright desolation.

Assassinate, assassinate

When tensions are running high, it only needs something minor to induce global calamity on a grand scale. History proves that the death of a single man can, in the right context, be sufficient to induce global catastrophe. The 1914 assassination of Archduke Franz Ferdinand of Austria-Hungary triggered repercussions that would guarantee the death of millions. Once again, the stage is set. The plan starts in Moscow and ends in meltdown.

Furthermore, we already have a working model for the potentially enormous planetary carnage the death of a single man can yield. Alexander Litvinenko. Relations between Britain and Russia deteriorated in tandem with the former KGB spy's health as he lay dying in a London hospital in late 2006. Scotland Yard and government officials were adamant that Litvinenko was secretly fed Polonium 210 in a sushi bar by Russian Secret Service FSB agent Andrei Lugovoi, possibly on the orders of the Russian state. Moscow insisted that Lugovoi had been framed by the British security services. Within months, Kremlin officials warned it was prepared to aim nuclear missiles at European cities for the first time since the fall of the Berlin Wall. Not to be outdone, British intelligence claimed that Russian spies in the UK were busy committing acts of espionage. It was all bubbling nicely out of hand when, at the beginning of the year, Russia's military chief of staff General Yuri Baluyevsky thrust his sallow, pudgy features towards the state cameras to make clear (as if anyone needed reminding) that Russia was getting a bit cross. Moscow, he declared, was prepared to use nuclear weapons. Pre-emptively. Against anyone. Bull's-eye.

All we now need is another Litvinenko-style death to light the touch paper. The victim must be Russian and their murder must occur in Moscow. Ideally, they should be linked to the FSB spy agency, although it doesn't have to be a senior figure. The murder must be mysterious. Think Georgi Markov, the Bulgarian dissident murdered in 1978 by being jabbed with a poisonous umbrella in London – although this time the Kremlin rather than the Houses of Parliament will frame the skyline.

Contacting a hitman is surprisingly straightforward. They can be tracked down through middle-men in Russia or through British-based expat communities. In a country where business is frequently settled by shady practices, rogue assassins are far from rare. A guaranteed hit, according to sources, should require no more than 500,000 roubles. Take care to pay in sterling. That way, the death can be traced back to Britain. Set up a false email account. Following the death, send oblique messages alleging that British security services ordered the murder. Try and include details that only an insider could know about the incident.

With the job done, sit back and watch the action unfold. Recent events indicate a certain inevitability to proceedings. As they follow the sterling and email trail, the Russian spy service will react with time-honoured aggression. A round of tit-for-tat follows, with expulsion of diplomats from respective embassies. Diplomatic relations between Whitehall and the Kremin collapse within days. As hostilities spiral out of control, a neurotic Russian media reports that Britain, backed by the US, is considering launching a military strike. Moscow ups the ante. General Baluyevsky, paler than ever, reiterates his threat of a pre-emptive nuclear strike. Except that, this time around, he specifically mentions the UK. He cites British-backed US plans to erect a missile defence shield across Eastern Europe as evidence that London's intention all along was to provoke Russia.

Once again, Gordon Brown insists his government played no part in what Russia has termed 'state-sponsored terrorism'. Later that day, in a briefing to international journalists, Whitehall officials make a welcome error. They remind the assembled hordes that Lugovoi has still to be brought to justice for the murder of Litvinenko. The briefing mistakenly suggests that renewed efforts to extradite Lugovoi for trial will be forthcoming. For those crammed inside the Kremlin, it is a tacit admission of a revenge murder. Russia goes the whole damn way. It presses the button. Whoosh. Several hundred thermonuclear warheads are unleashed towards Europe. Britain responds. On the few high streets that survive, sales of thermals and winter jackets (real fur-trim *de rigueur*, naturally) go through the roof.

If a job needs doing ...

Without wishing to spoil the party, a note of caution must be sounded. What if Russia resists pressing the button? What if the Kremlin loses its bottle, putting the planet and millions of its people before political expediency? Such a depressing scenario leads us to the DIY route. More 'atomic autumn' than 'nuclear winter', it is still worthy of consideration. The first step is to get hold of the ingredients, chiefly weapons-grade uranium or plutonium, and centrifuges which can manipulate normal uranium to the levels required for a nuclear bomb. Exact designs and centrifuge-production plans are known to be knocking around, thanks largely to a chap called Abdul Qadeer Khan.

Creator of Pakistan's atomic bomb, Khan is the best in the business. His network supplies the full caboodle – the nuclear materials, the technology, and the expertise to build a nuclear bomb from scratch. Don't worry that he is under house arrest in Islamabad; his huge clandestine network is still, according to intelligence documents, very much active. When, in 2004, he

RADIATING FURY

testified to selling nuclear weapons and materials to Iran and Libya, among others, many thought his empire was finished. But intelligence briefings reveal that 'most of Khan's accomplices remain free'. What is less well known is that at least four major figures in this shady network are thought to be British. Legal reasons prevent their identities being revealed, but they are super-rich and well known in the murky world of international arms smuggling. At least one once worked for the British government, licensing foreign arms deals, before being enticed by the profits to be made from dabbling in nukes. All four can be tracked down with minimal digging, their contact details available in the public domain.

But be careful. Approaches will be carefully vetted and potential partners must do more than re-mortgage. Facilitating nuclear war is a pricy business. One of Khan's associates was paid £2 million in commission just to help organize a deal. On the plus side, other bribes might not be required. Many countries have yet to introduce a law banning the trading of nuclear-related technology and few customs officers are briefed on what constitutes the technical equipment required to build an atomic bomb.

Let's go nuclear

Even if the DIY route fails and Russia holds steady, there is still ample cause for optimism. There have never been more states able to invoke man-made climate change overnight. Between twenty to thirty more are currently racing to develop a nuclear capability. Who knows quite what Iran is up to? Pakistan, meanwhile, is building a new heavy water reactor capable of producing enough material for up to fifty nukes a year. The UN's High Level Panel on Threats concludes that, rather than a reduction in atomic weaponry, the world is in fact embroiled in a 'cascade of proliferation'. If that were not cause for celebration in itself, the London-based International Institute for Strategic Studies concedes that attempting to control the trade in nuclear technology is a 'daunting task'. Russia may, after all, lose conviction, but sooner or later a man-made climate that doesn't rely on carbon-dioxide emissions will prevail.

WHAT'S THE DAMAGE?

* International agreements to curb the proliferation of nuclear weapons are not only honoured, but signed by all countries. **Not a prayer.**
* Rogue state acquires nuclear capability and strikes major world city. **Plausible.**
* Britain halts development of its new nuclear-weapon programme, Trident, saying it sets a shoddy example to the world. **Preposterous.**
* Regional nuclear conflict between Pakistan and India, or Taiwan and China destabilizes global ecosystem. **Distinct possibility.**
* US continues to build huge missile shield across Eastern Europe. Russia threatens a nuclear strike. It presses the button. **Conceivable.**

Likelihood of nuclear winter in next half-century: 69%

RADIATING FURY

25 Climate of fear

Dam it

AGENDA

* Raise a glass to world peace
* Draw up a draft for drought
* Net profit from a wet resource

There are three certainties in life: death, taxes, and the unerring ability of humans to make abundant resources scarce. Anyone looking for ways to unleash ecological Armageddon will appreciate the latter. Try removing access to water and see what happens. Analysis of half a millennium of human conflict – more than 8,000 wars – has concluded that water shortage is a profound trigger of upheaval. Experts identify 102 countries, together home to 3.7 billion people, where climate change and water-related crises could bubble over into violent conflict or instability. Depriving people of what they need, really need, gives you power.

Some 220 of the world's major rivers flow through more than one country. Each offers a sublime dynamic for possible conflict.

If the upstream state withholds its supply, people further down are gonna get tetchy. The planet is running out of water at a rate that, some believe, will soon make it the most valuable commodity on earth. Water is emerging as the biggest single security issue in the world. Welcome to the era of 'hydro-nationalism'. Water will be to the twenty-first century what oil was to the twentieth.

Nile: be seeing you

Just down from the temples of Karnak, locals sat staring at the eddying waters of the Nile. They had seen it low before, but this was getting ludicrous. Elsewhere, among the teeming streets of Cairo, the mood was fraught as the city coped with fresh rioting over high water prices. Without the Nile, Egypt would be finished. Upstream, Ethiopia was saying precisely the same thing. Yet, although Ethiopia owns the source of the Blue Nile, they aren't allowed to touch it. Not a pint-pot. 85 per cent of the water in Africa's greatest river surges from its soil, and all they can do is watch it ebb away into the arms of their angry neighbour. Only the truly perverse could have cooked up this little number. Britain of course. The Nile might meander through ten African countries but, almost eighty years ago, your colonial brothers had the foresight to broker a deal that, in your hands, will lead to one mighty scrap.

So, here is how it stands. Egypt wants to build new towns in the desert to sate a booming population. For this it desperately craves more water. Ethiopia desperately craves more water to stop its people dying of thirst. It is time for you to 'help'. Using a private company you must come to the assistance of the Ethiopian people. Discreetly, you will construct a massive dam using plans already prepared by the Ethiopian government. The dam will be sited just below Lake Tana in the Ethiopian highlands and very near to the river's source. Care must be taken to ensure that the construction cannot be spotted from the air; bank loans and aid must be carefully diverted to fund the project, but the

investment will be well worth it. With the dam in place, the hand is on the tap, braced for the signal to turn it off.

Overnight, the Nile will cease flowing through Egypt and Sudan. Billions of tonnes of water will back up behind the dam, forming one of the greatest lakes in eastern Africa. Ethiopia's 60,000-strong army will be deployed to its border. War will come quickly. Cairo makes no secret of the fact that any attempt to alter the Nile's status would be interpreted as an act of war. President Anwar Sadat could hardly have been more explicit when signing the 1979 peace accord with Israel, stating that his country would never go to war again except to protect its water resources.

In the aftermath of the dam, skirmishes soon erupt on the south-eastern Sudanese border, as troops attempt to invade Ethiopia. Meanwhile, in the north, frenetic fighting breaks out near Lake Nasser as Egyptian infantry attempt to head south through Sudan to reach the battlezone. Air strikes from Cairo engulf Ethiopia. Tanzania comes to Ethiopia's aid. Uganda gets involved. Kenya has a pop. Gradually, a chain reaction of violence ripples along the entire length of the Nile. By 2012, much of Africa is embroiled in its most vicious war yet.

Conjure profit from a dry spell

Not too far from Ethiopia, a corner of Sudan already offers a classic model for how water shortages can neatly cause conflict. A savvy combination of low rainfall and the advancing Sahara desert is blamed by many as the true genesis of the Darfur conflict. Many are in dire need of water but lack the billions of pounds required to build the infrastructure to transport and treat it. In your hands, water will become a recognized instrument of social control, as you cartelize access to man's most essential commodity. Act now. Soon everyone will be wanting a slice of the watery profits.

Of the many firms looking to control the world's water, only a few seem worthy of consideration. Lancashire-based Biwater might be intrigued by the Lake Tana project. Tanzania is among the ninety countries the water giant has been involved with. Britain's natty little archaic agreement with Egypt means that Tanzania is forbidden from doing anything that might affect the flow of the Nile. Consequently, the UK government backed Biwater to deliver clean water for Tanzania. For its goodwill, Biwater was set to make millions. But within two years, Tanzania's government cancelled the deal. The charge sheet of allegations was spectacular. No new domestic pipework had been installed, promised investment had stalled and water quality had declined. Biwater forcibly denied the claims and bit back, claiming that they had been misled from the start by Tanzania's water authorities. Obviously, with such experience, Biwater should remain in your thoughts.

Ideally, investment should be placed in companies with poor maintenance records and atrocious leakage rates. Similarly, those that manage to merge generous leakage rates with shameless profits should also be viewed accordingly. In this respect, Britain yields another two possible contenders. Thames Water (leakage: 894 million litres a day: profits £350 million) is looking to expand overseas. Severn Trent Water (leakage: 525 million litres a day: profits £150 million) is no slouch either.

Wat-er waste!

Of course, not everyone has money to fritter on water companies. But we can all do our bit. Pass a toilet, any toilet, and flush it. In one gush, the amount an African uses in an entire day for drinking, cooking, and washing will satisfactorily gurgle into the innards of the sewage system. Take a long, long shower. Better still, a bath. Take a two-week holiday, leaving the garden sprinklers on to ensure the lawn doesn't suffer in your absence.

In total, Britons consume a hundred times their weight in water every day. Make it two hundred. A thousand. Of course, you don't drink the stuff – not since they started to bottle it – but there's no excuse not to try and eat it. A modest 50-gram bag of salad from Africa requires almost 50 litres of water to produce. This 'virtual'

water trade is a neat way of plundering H_2O. Water is covertly stolen from where it is most precious. The perfect crime.

Irritation over irrigation

Recently, though, worrying signs have emerged that world leaders may be cottoning on to the psychotic behaviour produced in those who have no water. The issue has been raised at the top table of the UN security council. An internal Pentagon report confirms that dwindling resources will trigger 'offensive aggression'. Defence and environment ministers, meanwhile, concur that armed forces must start preparing for the inevitable round of 'water wars'.

For all that, you must remain grateful that there are still no internationally agreed rules on how nations should share rivers. Almost immediately after Labour regained power, more than ten years ago, a disturbing moment surfaced when the party publicly noted the potential for water conflict. In 1997, Labour sponsored a Watercourses Convention at the UN which, had it gone ahead, would have vastly reduced the prospect of future conflict. With remarkable prescience, though, they never bothered ratifying the convention in parliament. Recently, the then international development secretary Hilary Benn decided to explain what took place in parliament: 'We do not believe that any potential domestic benefits justify the resources that would be required.' Benn, the softly spoken child of firebrand father Tony, effectively told his colleagues to go ahead and f**k the planet. World peace, access to water, and ecological breakdown should never – and could never – be described as domestic concerns, said Hilary. Proof, again, that you can always count on the most unlikely of allies.

The non-clarity of international law is a cause for celebration. The level of potential for further conflicts remains reassuringly high. It will only get better. India's demand for water will exceed supply by 2020. In China, 550 of the 600 largest cities are running short. A world split between those who have water and those who don't is edging ever closer. Uganda's president Yoweri Museveni describes rising emissions as an 'act of aggression' by the rich nations against the poor. Well, come on then, Museveni. We're waiting. Museveni seems desperate to be remembered for his naivety. He once wanted Uganda to be powered by clean energy. Indeed, four-fifths of his country's energy once came from hydro-electric power, but now there is a drought. There is no water behind his vast dams. Poor Museveni, persecuted for his well-intentioned folly. In Egypt, where Moses once received the ten commandments, they wait for war. The rest of the continent waits for a biblical deluge. They wait in vain.

WHAT'S THE DAMAGE?

* By 2014, there are seven wars involving water shortages, each making Darfur look like a playground spat. **Probable.**
* The UN shelves targets to solve global water shortage. Two years later, people without safe drinking water climbs to 1.5 billion. **Likely.**
* Despite environmental pressure, Western Europe increases water use per head to around 160 litres a day. Bottled-water sales increase. **Certainty.**
* Ethiopia unveils plan to build dam on Blue Nile. Within two hours, Egypt launches air strikes on Addis Ababa. **Tenable.**
* In 2012 head of UN describes the threat of terrorism as far inferior to the security concerns from water shortages. The following month, Britons use a record 160 litres each. Another good year for power shower manufacturers. **Maybe.**

Likelihood of water conflicts by 2012: 84%

CLIMATE OF FEAR

26 Germ warfare

Vial bodies

AGENDA
* Get a lab job
* Spread the germ stockpile
* Embrace the open air
* Germ-inate, exterminate

With great skill and no little perseverance, scientists are busy preparing the germs that some day you plan to liberate upon an unsuspecting world. They might be about a thousand times smaller than the width of a human hair, but if these viruses are freed by the right hands they could reconfigure the planet as you know it. And some of your most favourite – Ebola, Aids, flu, and yellow fever – are just waiting for emancipation.

In a less than perfect world, these man-made viruses would be cocooned in impregnable laboratories by those petrified that their release into the environment could be sublimely catastrophic. It is with no little relief, therefore, that you have identified fifteen 'containment level-4' laboratories in Britain, laboratories that, while operating at the maximum biosecurity level, are certainly not impregnable and where the most infectious organisms are conveniently stored. Each lab handles some of the deadliest organisms known to man and nature: diseases that are highly contagious, fatal even in low doses, and impossible to treat.

As time goes on and research into bioweapons and other such hazardous organisms escalates, the risk of accidents increases with acceptable momentum. It strikes you as incredibly dangerous to keep all these germs in one place. Stockpiling is just asking for trouble. Being the dastardly cynic all this planetary meddling has inspired you to become, you deduce that a lot is going on behind closed doors. You must get behind those closed doors, enter the world of white coats and invaluable vials and disperse these germs for the good of your fellow men.

Put your foot in it

They can build tall walls, issue ID cards, erect automatic barriers, but you have identified the weak link in the world's biosecurity measures: people. Impressive regulations and safety protocols are rendered useless by carelessness. People get bored adhering to tedious procedures; researchers pick up bad habits or – shock horror – become complacent. They may end up washing contaminated material down the wrong sink or 'accidentally' removing equipment from the laboratory before it has been properly decontaminated. Perish the prospect.

It was a hot summer's day in 2007 when a damaging virus escaped from one of the country's most secure government laboratories, in Pirbright, Surrey. The facility managed to disseminate foot and mouth into the surrounding countryside. It was revealed that drains beneath the government-funded Institute for Animal Health laboratories carried waste, including the viruses responsible for animal diseases such as foot and mouth,

GERM WARFARE

bluetongue, swine fever, and, another longstanding favourite of yours, African horse sickness. Safety investigators subsequently discovered that leaking pipes, unsealed manholes, freak floods, and building work at the Pirbright laboratories had probably allowed the 01/BFS67 strain of foot and mouth virus to rampage around nearby fields and infect cattle. Result.

Infiltrate

To undertake your mission, first you must get access to the compounds. Get yourself a job at a laboratory whose work you believe involves pathogens deliciously capable of harming the environment. Initially, try one of the 350 'containment level-3' labs in the UK, some of them owned by industry, some by government, hospitals and universities. Academia might be your best in. The role of universities in overseeing security clearance for research students working with dangerous pathogens is currently under scrutiny by a parliamentary committee, betraying an annoying realization that there is an underlying problem but at the same time offering hope of a passport to potential pathogen paradise.

Professor George Griffin, chairman of the Advisory Committee on Dangerous Pathogens, has told MPs that he is perturbed by the lack of a national standard for people who work in high-security laboratories. Do not panic about not having relevant qualifications; training and risk assessment are the responsibility of the organization running the laboratory, and some, you suspect, will wave you inside – as long as you're cheap, and, let's face it, you're not in this for the money. Make sure you seem eager and affordable. So strapped of cash are university laboratories that Griffin admitted that they lack sufficient funds to adequately maintain storage for dangerous pathogens. Universities, explained Griffin, are forced to 'beg and scrape' for money to maintain high-security labs with insufficient funds 'for routine maintenance' at containment level-3 and level-4 facilities.

Of course, don't limit yourself to a university job just because they seem easy to snag. You are tempted by the riches on offer at the National Institute for Medical Research, where scientists study the deadly H5N1 avian flu virus, which has killed tens of millions of birds and spurred the culling of hundreds of millions others to stem its spread. You have learned that samples from infected people have been brought to the facility in north London for analysis. Equally exciting is information that researchers here have also been working on the 1918 pandemic flu strain that killed about fifty million people. If this sublime strain of the virus were allowed to make a fresh bid for freedom, it could cause a new pandemic. Virtually no one would have immunity. As humanity struggled with its latest threat, any care for eco-living and dolphins would be first to go out of the window. Normally just nine members of staff at a time are cleared to work inside this lab, so put on your best tie. CVs will certainly need embellishing for this one, not to mention the perennial staple of the bogus reference.

The NIMR facility is such a tempting target there's no wonder it is one of the so-called Big Five, those authorized to handle the most dangerous diseases. The MoD's Defence Science and Technology Laboratory, at Porton Down, on Salisbury Plain, Wiltshire, is another and carries out high-level research on diseases such as anthrax and bubonic plague. Checks here are, however, tiresomely strict, and instinct suggests your efforts are better placed elsewhere. Try the Health Protection Agency, which has its Centre for Emergency Preparedness and Response on the same site and runs a similar lab at the Centre for Infections in Colindale, north London. The SARS virus, Lassa fever, and the Ebola virus, which causes massive bleeding in victims, are handled there by those privileged enough to enter its inner sanctum.

But your most desired target may have yet to arrive. Plans are afoot to build a laboratory housing dangerous pathogens in central London. The government-backed £500 million UK Centre for Medical Research and Innovation is scheduled to be built in St Pancras and, at this stage, is likely to include work on the usual

GERM WARFARE

selection of hazardous agents. Daunting as some of these laboratories might seem now, do not be dissuaded; this is one plan where keeping your eyes on the prize is vital. When in doubt, think back to 2005, when a pandemic strain of Asian flu was released by an American laboratory after it was accidentally put into test kits sent to scientists around the world. Probably your favourite 'accident', though, occurred in the former Soviet Union in 1979, when weaponized anthrax bacteria escaped from a bioweapons facility in the Urals. Scores died when workers changing shifts left a vent unfitted with a filter, allowing the germs to escape.

Disseminate

Once you are inside your secure laboratory, stay cool. Put on your white suit, comb your hair, learn the lab layout by heart and, in particular, where the most harmful vials are stored. Each lab has different protocols and, initially, it is important only to blend in. Soon enough, you will be viewed as hard-working and trustworthy. You become known for your diligence in removing all clothing before entering and showering before leaving. Eventually, you will be handed the keys to the laboratory's most prized contents. You will probably decide to smuggle stuff out one night after working late. By then, you will have taken the premises' most dangerous airborne pathogen. Although viruses normally require a host, into which they inject their genes and hijack the cell's biological machinery in order to duplicate, some can survive in the atmosphere for weeks. Foot and mouth sets an inspiring example, a virus which in 2001 blighted the UK economy and rural regions of Northumberland, Cumbria, and the West Country. But – and this shows the real power of a carefully targeted viral release – it has also had, and still does, fabulous international effects. With British meat effectively off limits after the 2001 foot-and-mouth epidemic, Brazil stepped in to make up the shortfall in meat production and began razing the Amazon forest like never before so it could plonk massive cattle ranches where virginal rainforest once stood.

Finally, if you fail to land a job, do not completely despair. In that scenario, contact one of your like-minded doctor academic friends. Doctors and scientists, as the UK terrorist attacks on Glasgow airport a year after Pirbright's foot-and-mouth release proved, can be radicalized like the rest of us. But be selfish, really the prize is yours for the taking, and soon your white-coated spine will be shaking with the greatest viral release in humanity's history.

WHAT'S THE DAMAGE?

* New international standards announced for containing dangerous pathogens. **Certain.**
* New lab opens in central London, but causes commuter chaos with bi-weekly closure of main London rail lines due to scares about the facility's bio safe zone. **Predictable.**
* Lethal pathogen escapes from university lab. Cover-up revealed twelve months later by politicians suffering unusual symptoms of conscience and public concern. **Possible.**
* Bio-terrorists strike National Institute for Medical Research. They gain access to avian flu stocks, but are forced to surrender after being surrounded by armed police. **Maybe.**
* Protest groups reveal a number of previously serious but unreported incidents at secure government laboratories after freedom-of-information documents released. **Likely.**

Likelihood of dangerous virus release by 2015: 68%

GERM WARFARE

27 Not so slick

Transports of joy

....

AGENDA

* Go with the flow
* Become an oil baron
* Claim your petrol on expenses
* Wriggle out of polluter-pays fines

In this age of the Anthropocene, the destructive Era of Man, there is no more representative image than a seagull smothered in oil. Robbed of flight by the weight of tar, one can barely imagine what must be going through the bird's tiny mind. The world's economy relies on such a charming substance, with 81 billion barrels produced each day. Moving the stuff about is a fraught and lengthy business, full of opportunities for accidents. A tanker takes nearly forty-two days to travel from Saudi to Texas. The short hop from England to Holland takes two: it would be quicker swimming across with each barrel. Oil spills happen with wonderful regularity, but you need only pay attention to the biggies. They will help you to put the planet to bed. Whether you employ terrorism, the advancing decrepitude of freight transport or rely on Mother Nature herself, the massive oil slick belongs as much to the future as to the past. And, best of all, you'll sail into the sunset scot free

Slick bitch

Finally, just after 8 a.m. on 19th November 2002 the great hulk of the tanker *Prestige* juddered, then split in two. Even before she sank, the Atlantic was turning black as twenty million gallons of oil leached from her broken form. But that was just the start. For weeks, the *Prestige* dutifully bled 125 tonnes of oil every day. Environmentalists wept pathetically from the nearby Galician coastline as the dark tide engulfed 350 miles of its ecologically fragile shores. They had reason to cry. Knowing exactly how to maximize the effects of this toxic cargo, the tides had dragged the oil towards the treasured coral reefs of Galicia. Even without your intervention it would take years to clean; if you joined the efforts, it would be a never-ending chore.

In total, 64,000 tonnes of fuel oil joined the tide. The *Prestige* is not only a lesson in what is achievable when one of the rust-buckets of the tanker world sets sail weighed down with pollutants, but a motivational reminder that even the most high-profile environmental disasters can be executed with impunity. Six years on, and no one has been prosecuted for the lavish pollution caused by the *Prestige*. A detailed examination into the events that led to her falling apart off the coast of Spain confirms that it is very much possible to get away with anything behind the opaque mesh of shipping ownership, corruption, and lack of regulation. A broken trail of evidence means that no one really knows who owned the oil that washed up on Galicia.

Pliable liability

Without further ado, you must concoct a repeat, using the *Prestige* disaster as a template for success. The ship was owned

NOT SO SLICK

by a Liberian front company called Mare International. Their accident insurance on the vessel was £15 million, a mere hundredth of the clean-up costs of Spain's Galician coast alone. Mare International appeared to be owned by a secretive Greek shipping dynasty, the Coulouthros family, which in turn operated under a company called Universe Maritime. But the *Prestige* sank with the flag of the Bahamas flying from its decks after it was registered with the Bahamas Maritime Authority, an organization that may conjure images of palm trees and all things tropical, but in actual fact has its offices in the City of London. So far, so complicated. The sludge that belched from the *Prestige* was being moved across the planet by an oil trading company called Crown Resources which, although formed in Gibraltar, has its headquarters in Zurich, but also works from a major office in an exclusive address in London's West End. At the time of the disaster, at least five of its directors were British.

Back in the summer of 2002, word reached ship broker Stefan Giesen, of the brokerage firm Petriam, that Crown Resources was on the look-out for a ship to shift some 70,000 tonnes of oil across the planet. He had found the ideal vessel. The *Prestige* had recently been given a clean bill of health by US shipping authorities. Crown would take it, hiring the *Prestige* for £13,000 a day. For Crown, which traded millions of tonnes of crude oil across the globe, it was just another anodyne deal. Nine days earlier, when the *Prestige* anchored outside Gibraltar to refuel, British authorities had not seen fit to inspect the single-hulled tanker.

Meanwhile, in Russia, news of the deal would have undoubtedly pleased one of the country's most prolific entrepreneurs, Mikhail Fridman. This oilman had founded the Alfa Group Consortium, a powerful Russian conglomerate which happened to own, among a myriad of interests, Crown Resources. At the time of the catastrophe, Fridman was the ninth richest man aged under 40 in the world and worth more than a billion pounds. Now, aged 43, he has almost £7 billion in the bank and has deservedly nudged into the top fifty richest characters in the world. Among other companies listed within his consortium was Tyumen Oil (TNK), a relatively new oil business. Four months before the spill, the company had been investigated but officials found no grounds for concern. Even so, misgivings were voiced by the European Bank of Reconstruction and Development over the deal that had led to TNK acquiring the massive Siberian Samotlor oil field.

Samotlor is an absolute gem. It is probably among your favourite places on the planet. Lakes of oil are regularly reported on the tundra surface. An independent study published a year before the *Prestige* sank revealed that up to 2 million acres of land there were polluted by oil. Rivers and underground aquifers were contaminated by up to fifty times the Russian safety standards, which are hardly the most stringent in the world. Samples of drinking water taken over five years showed that 97 per cent were poisoned with oil. One report estimates that pipelines in the region were leaking 500 litres of oil every second. Some of Britain's biggest names in business pricked up their ears – such generous leakage rates meant a whole lot of oil was still to be had. Weeks before the *Prestige* sank, City grandees Sir Peter Walters and Sir William Purves joined Tyumen's 'supervisory board'. Purves is a former director of the Shell Transport and Trading Company. Sir Peter is a former chairman of BP, which had just proudly announced world-beating environmental credentials.

Less than a year after the *Prestige* sank, and with its toxic cargo still lapping the shores of Europe, BP merged with TNK in a £3.5 billion deal to operate in one of the most immaculately polluted and environmentally damaged oil-producing areas on the planet. Samotlor. In more ways than one it was a shrewd move. By 2007, TNK-BP was producing 1.8 million barrels of oil daily.

Soon enough, James Harmon wanted a slice of the action. Harmon joined the TNK advisory board, having left his position as chairman of the US Export-Import Bank of the United States. While he was chairman, the government agency had approved the

NOT SO SLICK

loan of £250 million in credit to TNK for the refurbishment of the Samotlor oil field. It was so controversial that the White House had tried to block the deal. They were worried about the Alfa Group's alleged Mafia connections. It mattered not. Houston-based energy company Halliburton wanted the deal to go ahead. And they were pretty well connected. Leading from the top was their chief executive. He was called Dick Cheney. So good was he that Halliburton was awarded £146 million for the refurbishment of the Samotlor field and Cheney went on to become the second most powerful figure in the world, the US vice-president.

The truth is that the oil which inked the southern and western coasts of Europe perhaps came from Samotlor, but no one really knows. Crown traded significantly with TNK Tyumen, importing up to £45 million of Russian crude oil into Western Europe each month, but there the trail ends. Possibly it came from Iraq. Six weeks before the *Prestige* slick, Alfa signed one of the largest deals in the corrupt United Nations oil-for-food programme, taking twenty million barrels a year from the Middle Eastern state.

Since the *Prestige* went down, oil and shipping industries have continued to steer clear of international controls and regulations. The search for a suspect has foundered. Attempts by Spain to prosecute US shipping bodies for having declared the *Prestige* effectively seaworthy have collapsed. And the plot gets murkier. It emerged that, shortly before it halved, the ship had been inspected and surveyed in St Petersburg, Dubai and also Guangzhou, China, where repairs were made to the part that subsequently broke.

Culpability, Brown?

Currently, the EU is attempting to force through a law on the 'polluter pays' principle, which would somewhat take the edge off trying for a repeat. But industry lobbying has forced through a clause which means the law will only apply to protected habitats, as well as numerous other exemptions. Subsequent negotiations have established the International Oil Spill Pollution

Compensation (IOPC) Fund, made up largely from levies on oil companies, and which could pay up to £90 million compensation for damages, if the trails weren't so hard to follow.

Meanwhile, more than four years on, the *Prestige* continues to do admirable work. New slicks have been detected, up to 23,000 tonnes of oil remain in the ship's carcass, and bacteria corroding the hull could soon produce a rupture. Over the last forty years, major oil spills have occurred every two years. Some time soon, a shipping broker will say he has the perfect ship. Soon afterwards, a supertanker carrying 250,000 tonnes of oil will rupture. And you'll be there, if required, to help them cover their tracks.

....

WHAT'S THE DAMAGE?

* Spanish government announces they know who is to blame for the *Prestige* disaster. **Never.**
* Massive supertanker accidentally beaches below the white cliffs of Dover. Channel turns black. Ship's captain never heard of again. **Imaginable.**
* 'Polluter pays' legislation forced through, far stricter than currently proposed. For each drop of oil spilt, an immediate £100 fine. **Not a chance.**
* In 2017, the *Prestige* surrenders the remainder of her cargo. On the same day it emerges that a gigantic spill at the Samotlor field has been covered up for seven years. **Possible.**
* Terrorists sabotage a Shell oil tanker, polluting majority of Niger Delta. Shell refuses to pay clean-up costs, declaring legal action against alleged attackers. **Extremely likely.**

Likelihood of oil supertanker sinking and causing huge environmental disaster by 2015: 84%

....

NOT SO SLICK

Politically Incorrect

People versus planet

Grin and beer it

28

Bottoms up

AGENDA

* Eat, drink, and be merry
* Raise the parting glass
* Go out on a high

INVITATION

Dear Fellow Earthf**k,   *21st June 2011*

Latest reports confirm that the wretched planet has finally passed its tipping point. Time then for a knees-up. As a small gesture commending your spirit and fortitude, we would count it an honour if you were able to attend our forthcoming end-of-the-world party, The Big Fat Carbon Bash. Private flights only.

Yours lovingly,
M. Townsend & D. Glick

Feteful event

Your invitations are sent by airmail. There were two important factors to consider when choosing the venue; first, that it was very remote and, second, that it was ecologically delicate. You have chosen a small atoll in the Maldives. With sea levels rising by 0.9 metres a year, this string of islands should have drowned long ago and with the help of your party it will surely be among the first places to disappear through climate change. The best way to reach your island is by private Learjet. One flight across the US is equivalent to driving a Hummer for a year – and some of your guests will make journeys four times that distance.

The party itself will be an affair to remember, a carbon-intensive all-nighter. Naturally, it will be staged in a draughty marquee where, due to the sea breezes, huge stand-alone industrial-sized patio heaters will keep guests warm throughout the night. Even if it is a typically humid night, you'll want to keep them turned on – you've paid to rent them after all. Massive generators are being flown in from 5,000 miles away to power your impressively large lighting rigs, which will prevent any of your guests wandering off to stargaze by radiating glare into the heavens above. Of course there must be music, from a sound system capable of waking up the whole Indian Ocean and affecting the mating and migration habits of any poor brute – winged, footed, or finned – who happens to be lingering in this part of the world. You plump for DJs Dillinia and Lemon D, creators of the mesmerizing Valve Sound System. Their touring sound system thumps out drum 'n' bass with a total power output of 96 kilowatts – several dozen times the volume of normal nightclub systems. The noise might even shake the Indian mainland, 500 miles away. Your recce of the area found that the best spots for the DJ booth happen to sit next to a sensitive nesting site and an idyllic cove hitherto undisturbed by man or time. Either would offer beautiful surroundings for your guests, saving you on the cost of decorations (although you have budgeted for balloons – latex, naturally, which will end up deflated and discarded about the island's fragile ecosystem. These balloons have been recognized for their efforts in killing marine wildlife in England. No doubt they'll behave just the same abroad). On one of these sites you'll position fifty-two subwoofers, arrayed in six enormous speaker stacks. The whole system is transported in three 7.5 tonne lorries so you've requested twelve private flights to bring it in. In the past, environmental-health authorities have stopped this impressive hi-fi from being played at full power. You don't intend to worry about regulations: there are no officials on the guest list and your bouncer won't be letting any greenies across the threshold.

Menu for disaster
A formal menu will serve up some of the world's favourite foods.

~ *Starter* ~

Champagne and Caviar
A 2004 vintage from the Champagne-Ardennes region. Grape harvests were lovingly tended with prodigious lashings of nitrogenous fertilizers. Caviar sourced from unsustainable stocks of wild sturgeon, now protected under international regulation, delivered from the Kazakhstan black market direct to your paper plate (which you are encouraged to frisbee into the Indian Ocean immediately upon completion). Salut.

~ *or* ~

Prawn Cocktail
Flown in from industrialized Bangladeshi shrimp farms, linked to the destruction of ancient mangrove swamps. Some trawled from the South China Sea where, for every kilo of shrimp served tonight, another 40 kilos of marine animals have been discarded. Sources say that 150,000 turtles are killed in this way. Enjoy.

GRIN AND BEER IT

~ or ~

Shark-fin Soup

A delicious broth made using fins from the North Pacific scalloped hammerhead, a species which has declined by 98 per cent in recent years. Recently classified as endangered, rest assured your fin has been sourced from illegal traders who scythe fins straight from the torso and leave the live carcass to drown.

~ Main Course ~

Steak Tartare on bed of Rocket

Brazilian steak from soya-munching cattle in a freshly-cleared quarter of the Amazon rainforest basin. Served with battery-farmed eggs from Honolulu on a bed of pesticide-soaked rocket, with side of English potatoes sprayed no fewer than eighteen times with various chemical treatments.

~ Dessert ~

Dead Forest Gateau

Two-metre-wide spherical cake modelled on the globe and encircled by flaming torches. Made from processed sugar and no less than seventeen artificial colourings and additives. Sprinkled with Ghanaian cocoa, the cultivation of which has destroyed several unspoilt habitats and poisoned nearby water supplies.

Can-do spirit

You ordered the most important ingredient well in advance: 2,000 cans of Miller Lite, chosen tipple of climate-change deniers everywhere, brewed by massive international brewing company SAB Miller, of which a substantial shareholding is owned by Altria.

In turn, Altria is owned by the very company that initiated corporate funding of climate-change denial. Yes, you guessed it: the world's largest tobacco company, Philip Morris.

These Millers taste good out here in the wilds of the Indian Ocean, you think, as you down your seventh can, pausing only to wish the manufacturers hadn't dispensed with the ring-pull, a handy invention that transformed the can into a drink-on-the-move must-have, never mind choking small creatures and lacerating animals' feet. You remember that you still have the plastic four-pack packaging, which is brilliantly invisible in water and deftly guarantees that wildlife chokes or becomes ensnared. You toss the can into a shallow-water coral and head back to the huge row of refrigerators bulging with climate-change-denying booze. Of the UK's carbon emissions, 1.5 per cent is estimated to be due to the cooling of lager, cider and white wine and, as you pull out another tinnie, you wonder just how far it travelled to reach here, bearing in mind that a bottle from one of the four main UK brewers accumulates 24,000 miles in production and transport.

To make a single can requires the same amount of energy needed to power a television for three hours. In fact, every tonne of aluminium processed from primary ore emits 1.7 tonnes of carbon dioxide and an additional 2 tonnes of carbon dioxide from perfluorocarbons, potent greenhouse gases that linger in the atmosphere for centuries – although they probably won't get the chance to prove more than a fraction of their longevity.

Rio Tinto is one of the biggest producers of aluminium cans; the massive electrical smelters they use suck up as much energy as a city of one million people. Aluminium smelting requires more energy than any other metal process and Rio Tinto still rely principally on coal. They obviously believe unerringly in the power of inefficiency. Their vast smelting plant in Gladstone, Australia, actually makes useful energy from only a third of their coal supplies, while enough carbon dioxide is produced in making each can to fill three hundred of the things.

GRIN AND BEER IT

Get wrecked recklessly

Most of your invitations have been sent to boozy Britons, partly because they are guaranteed to drink far too much. While the Germans also like a drink, they are far too prone to recycle. Britons just cannot be bothered. While the average UK household guzzles the contents of 208 cans every year, they recycle just 41 per cent, less than half of their fellow German drinkers' efforts. More than 3.5 billion cans a year are discarded in fields, landfill sites, and on pavements. The energy required to smelt down aluminium every year is enough to power an episode of *EastEnders* every night of the year in every country across the planet, which is probably the most cogent reason anyone has formed to avoid recycling cans.

You want the party to carry on through the night, ideally until the sea levels around your Maldivian idyll rise and rinse the remaining party-goers from the dance floor. To guarantee you'll stay awake you will need a helping hand, and there is probably no better way than the old conk-dust. You will source the cocaine as directly as possible from the FARC guerrilla group, whose bloody campaign of attrition, murder, hostage-taking, and extortion is funded directly from the profits from the coca leaf. Once a left-wing revolutionary movement, FARC is now the world's leading trafficker in cocaine and has long been suspected of running the Colombian cocaine industry. The oldest functioning guerrilla organization in the world, it is richer, more numerous, and better armed than any other single Colombian drug cartel.

FARC operates with Venezuelan state officials on the ground, where military and drug-trafficking activities coincide. 30 per cent of the 600 tonnes of cocaine smuggled from Colombia each year go through Venezuela. You can also bank on the help of the Ecuadorian authorities to facilitate the arrival of your stash.

Time to place your order. The price has halved over the last decade and, currently, cocaine fetches £40 a gram. Don't take any notice of that. Tell your contact you are prepared to pay top-whack London prices. Order a kilo – that's ten grams per party-goer – with a 95 per cent profit margin to be cut among the myriad of dealers along the line back to the FARC guerrillas. Anything for a cause! Lots of cocaine is used by people who claim to be ethical; you might consider inviting Alex James, the Blur bassist who ingested £1 million worth of booze and coke but is a clean-cut cheese farmer these days. Tell him that it might be the last party he's invited to. Tell him he can bring one of his goats. Your order will eventually filter back to the remote airfields in Colombia from where planes fly to Venezuelan airfields. Intelligence sources indicate that they then continue on to Haiti or the Dominican Republic, from where you will commandeer a special flight to the Maldives, a cool 9,000 miles to the final and best party the planet will ever record. At last, let's call time on all tomorrow's parties. The hangover won't be pretty ...

....

WHAT'S THE DAMAGE?

* The Maldives sink below rising tide by 2015. Final act of island government is attempt to sue Western states for driving climate change. **Possible.**
* FARC is embroiled in escalating war with Colombian government. Thousands die in the cocaine wars. Use continues to increase throughout Europe. **Likely.**
* Miller becomes one of the most popular beverages in the world. Alcohol consumption continues to climb in Britain. **Almost certain.**
* Plastic four-pack can packaging is banned after last surviving panda chokes to death one night in zoo enclosure. **Plausible.**
* As the planet collapses, people grow more reckless and fun-loving as hope all but fades. **Likely.**

Likelihood of end of the world party by 2020: 27%

29 When Porsche comes to shove

Put your foot down

....

AGENDA

* Drive to extinction
* Accelerate for change
* Floor it

Foot to the floor, as fast as you can. Your Porsche Cayenne is a metal-boxed version of heaven. Some may say you are phallically challenged because you have invested £70,000 on a car that rarely travels more than 40 miles per hour, and that's on a good day. But you merrily bend the law, knowing that in the process you are helping to exacerbate the meteorological vagaries of climate change. The greens believe that, along with the atomic weapon, the combustion-fired engine is the Devil, but the thrill of speed itself has never murdered anyone; it's the sudden stopping that hurts.

The greens can get on their bikes, so long as they keep to their cycle lanes; you have a fast car to drive.

Contest the congestion charge

It's April 2008, and you are fighting for your right to pollute. Porsche, bless their sleek metallic souls, are dragging the mayor of London to court after he proposed a £25 a day charge (up from £8) to drive the zippiest, sleekest cars around the capital's centre. Ever willing to help the cause, you log on to their website and pledge your support. The Porsche petition website features a blurred wonder shown whizzing around Parliament Square and is raising online signatures from motor-mouthed people across the planet. At last, you think, a motor manufacturer with the guts to stand up and fight against the bully-boy tactics of the green lobby. In your less reasonable moments, you believe this tax to be the most unfair, unreasonable, and disproportionate levy ever. Your motoring friends claim that, per tonne of extra emissions, this tax is 3,500 times as much as people should be paying. While your Cayenne Turbo is one of the most polluting vehicles around (why else fork out the cash?), with an above-reasonable 605 grams of carbon dioxide released per mile, you suspect the tax is more about squeezing money out of the motorist than saving the planet. The former mayor reckoned that higher charges would trim a pathetic 5,000 tonnes of CO_2 each year, with the money raised funding things such as cycling. You note that pollution levels have not changed since the introduction of the congestion speed limit and, as for congestion – well, take a look outside. Recent figures show average traffic speeds in central London are now lower than before the congestion charge was introduced in February 2003. And, with hundreds of cars set to be given free access to the city, there are fears the jams will get worse. London will be packed with cars full of people with limited style and non-existent social skills, and what sort of message is that to be sending from a so-called world city? While they get away

WHEN PORSCHE COMES TO SHOVE

scot-free, it will be the richest, the most successful, the most intelligent people – those who, frankly, give the city its image – who will be overcharged.

You are furious with the mayor and, in protest, have written a personal letter of solidarity to Wendelin Wiedeking. You have always been a fan of Porsche's chief executive, who, after all, does earn £55 million per annum and so therefore knows what it takes to be successful and drive decent cars. Wiedeking has said: 'I recognize that a "fair deal for Porsche owners" is unlikely to be a rallying cry that will see millions marching on the streets of London.' But he is wrong. Everyone you admire respects Porsche and its heroic fight to try and help you drive high-performance vehicles for as long as you can afford them. Let it not be forgotten that the German car-maker was pivotal in resisting attempts to tackle the car industry's contribution to climate change. In fact, along with countrymen BMW and Daimler, Porsche was involved in lobbying to reduce and delay Europe's mandatory targets to reduce carbon emissions from cars. However, despite the painfully predictable evidence presented by scientists, the European Commission listened to the more impressive voice and opted for a blueprint on emissions limits that avoided hurting car-makers more than was necessary.

Wedded to Wiedeking

Back in 1994, a German environment minister advocated a limit of 120 grams per kilometre (about 190 grams per mile) as the maximum amount of carbon dioxide that should be released by the average car. This move would have meant a 3 per cent reduction on emission levels at the time. Almost fifteen years on, it was his country that also led the way in abandoning such ludicrous targets. Despite statements from leading European politicians that the 120-gram threshold was essential, Germany's car-makers pushed for a lower limit, and the EU's executive went on to recommend a significantly less ambitious target of 130 grams per kilometre (around 210 grams per mile). Porsche had done it again. It was a landmark victory.

You remember the emotional warnings from German car bosses of vast job losses if they were forced to shed the extra 10g – according to Wiedeking, the guidelines represented an 'attack' on the German car industry. No convincing evidence was produced to show that the calamities predicted by scientists were realistic. Thus Wiedeking was able to lambast Europe's plans as 'wholly alien' and to state that they breached the laws of physics. He never mentioned the almost 350 grams per kilometre produced by your beloved Cayenne, because he never had to. He had won. Between 1990 and 2005, a fall in greenhouse-gas emissions was recorded in almost every economic sector in the EU apart from transport, which climbed by more than 30 per cent, with cars and vans accounting for about half that increase. This was also brilliantly glossed over. The industry continued, unfettered, to produce the streamlined machinery that will speed the planet to its fate.

Out with the old

If friends do get hoity-toity about the ethics of becoming speed-freaks in a slow city, then by all means persuade them to get a new eco-car. The very manufacture of a car constitutes a big part of its carbon footprint. Remember when Vauxhall promised to give a £1,000 trade-in for your old banger, regardless of age and condition, in the name of the planet? While it is true that, often, a decrepit, badly maintained vehicle will emit more pollutants from its exhaust pipe than a new vehicle, this ignores a key fact – namely the several tonnes of carbon dioxide that are produced in the manufacture of a new car and the disposal of an ancient one. But you might want to shun anyone who opts for the ghastly Toyota Prius, which has become the hybrid petrol-electric car of choice for eco-conscious motorists. Although Toyota were pulled up for exaggerating claims about the carbon emissions of the Prius, their popularity continues. It's embarrassing really. For the ecocide

convert, telling anyone that you drive a Prius is like telling someone you are about to kiss that you have glandular fever.

You look past your framed portrait of Wiedeking and out through the window. Below is the Victoria Embankment, London, a riverside drag-strip where City lawyers in Cayennes vie with motorcycle couriers to see who can burn the most rubber in the 100 metres to the next traffic light. You hope that there will never be an end to such scenes. You gaze at Wiedeking and wink.

WHAT'S THE DAMAGE?

* Campaign against fast cars and 4x4s sparks mass slashing of tyres throughout Europe. **Probable.**
* Congestion charges of £40 for all high-polluting vehicles introduced in every major European city by 2015. **Unlikely.**
* The driver of a Porsche Cayenne is dragged from their vehicle in Mayfair by eco-activists. The car is torched and the driver covered in green paint and tar. **Possible.**
* Sales of eco-friendly cars double those of gas-guzzlers by 2014. **Likely.**
* By 2015, every prominent European politician has acquired a Prius in which to drive his family to the airport for an exotic long-haul holiday. **Odds on.**

Likelihood of gaz-guzzlers becoming commercially extinct by 2015: 41%

30

Greenwash

Talk the talk

AGENDA

* Green up your mission statement
* Encourage green scepticism
* Wise up to jargon
* Make unfettered money

BAE Systems, one of the world's biggest arms manufacturers, had an image problem. Their bullets had maimed children and killed untold innocents. The company executives convened to come up with a somewhat softer image. The answer, they agreed, was simple. They began developing green bullets. Not bullets in shades of jade, lime and turquoise – no, environmentally friendly bullets. Their new range of lethal projectiles offered clear advantages over the ghastly traditional type, which, BAE revealed, could 'harm the environment and pose a risk to people'.

Business, BAE bosses believed, would boom. More bullets would be sold. Environmentalists would rejoice as conflicts were settled with ecologically aware killing machines. BAE executives knew what they were doing when unveiling a bullet tipped with tungsten rather than lead. They were merely following one of the new tools of business. Throw in a 'green' here, an 'eco' there and, hey presto, your arms-manufacturing giant is a friend of the planet.

Paint it green

Words are powerful things. People guzzle gallons of cola, just because it's 'lite'. Treated sewage sludge is rebranded as 'biosolids' and suddenly the world is clamouring to spread it on their soil and crops. The beauty of greenwash is that the consumer justifies excessive spending with the belief that they are saving the planet. Hogwash. In actual fact, excessive consumption can only ever be a threat to the biosphere.

But encouraging people to overconsume is only part of the greenwash genius. Businesses who wouldn't recognize corporate responsibility if it came up and booted them in the behind cleverly exploit environmental credentials as a trendy means to boost profits. Over time, consumers will become flannelled to distraction by an overkill of good intentions. Ultimately, the words 'eco', 'sustainable', and 'corporate responsibility' will come to mean nothing. Cynicism will mount. Even the truly environmentally friendly companies will be mistrusted, and everybody will go back to buying the usual old environmentally harmful crap.

In the golden age of ostentation and heady consumption, 'greed was god'. Now, green is god. PR puffery, idle claims of eco-virtue, and fake promises are the *modus operandi*. Once, people feared that protecting the environment came at a cost to business. In actual fact, claims of environmental virtue are a boon to business. People will buy anything if they believe it is nice to flowers and dolphins. But of course it is easier and less expensive to change the way people *think* about reality than it is to change reality. All you need do is to *say* you are protecting the environment. No action is necessary. Greenwash is the new greenback. For now.

Oil the wheels

When BAE announced its green bullets, it was the final proof that greenwash had taken over the world. This shameless green behaviour had already been adopted by numerous other companies. Just hours before the 2007 United Nations conference on tackling climate change, oil conglomerate Shell found itself talking about how it was committed to a low carbon future. Five days later, tucked away in a press notice cunningly designed never to be, er, noticed was news that Singapore-based Environ Energy Global had bought Shell's solar photovoltaic operations in India and Sri Lanka for an undisclosed sum. Solar energy no longer appeared central to Shell's 'low carbon future'. They were following the greenwash code to the letter – in fact, they were adding some of the letters and taking others away and, hey, almost rewriting the rules. The previous year, also buried away in an unheralded announcement, Shell had sold off its solar-module production business to a German firm. Greenwash doesn't always have to be fancy words. Being economical with the truth is equally effective.

At the same time as Shell was banging on about low-carbon dreams, word was spreading about the ambitions of another oil bedfellow. BP had long been famous for its Beyond Petroleum campaign. Beyond Parody might have been a more apt title. The company was simultaneously gearing up to invest £1.5 billion in mining Canada's tar sands for oil. They would be the world's dirtiest oil mines. Campaigners describe the act as the 'greatest environmental crime in history'.

There is much to learn from BP and Shell. Both are among the biggest producers of greenhouse gases in the world. Both claim to be 'environmentally friendly'. Shell's adverts for a greener future regularly pop up on TV, and inevitably they sometimes go too far. One advert in particular showed what can be done if you just have the balls to go for it. Against a soothing soundtrack and swirl of Technicolor petals, it revealed that the oil company uses its waste carbon dioxide to grow flowers. Technically true, but just

GREENWASH

0.325 per cent is used for that purpose. Unfortunately, in this instance, the non-technical truth outed and the advert was promptly withdrawn.

Don't call a spade a spade

To be a greenwash expert, a sophisticated balancing act is required to hide the truth without resorting to outright lies. For a reasonable £500 an hour, a host of PR companies will dream up ways to disguise the actualité. Advertising a green weekend break? Stick an energy-saving lightbulb in the all-night bar. Install a dimmer switch above the jacuzzi. One Devon hotel found that having 'drought resistant hanging baskets' was sufficient to label itself green. The world's biggest diversified mining group, BHP Billiton, promised to 'find lasting solutions consistent with our goal of zero harm'. Zero harm involved reducing accidents to its workers, not a pledge to stop scooping out the innards of the earth. But it sounded good. Here are six simple steps to follow if you want to abuse first the truth, then the planet.

1. Create a perfect environmental image. Park a gas-guzzling SUV beside a pristine stream. Sunlight glints off the polished twin exhausts. A doe-eyed deer nuzzles its bull bar. Click. Capture the image and place it in a glossy magazine read by people with too much money. They will associate gratifyingly wasteful 4x4s with the charms of Mother Earth.
2. Be selective with details. List charities your benevolent company has supported. Don't be so vulgar, though, as to cite actual amounts. Philanthropists don't feel the need to scream about their generosity and neither should you. Be sure to mention your donation to save the rare marshes in Mozambique. Omit to reveal that your company offered a one-off payment of £5.
3. Employ the distraction technique. Your firm makes bad stuff called acetic anhydride. Invest modestly in an organic farm and publicize accordingly. The environmental arm of the business is

an 'ardent supporter of local produce'. Just be careful not to mention, erm, acetic anhydride.

4. Choose your words carefully. Fresh. Clean. Eco. Friendly. Pure. Your company's mission statement should include at least one of these choice words. Even better, modify your firm's name. Grimshawl becomes Ecoshawl. PreTex becomes enviroTex. Be opaque with subsidiary companies. Pharmaceutical giant Procter & Gamble created a Future Friendly label. Shoppers had no idea P&G were behind it. What did it matter? It was a sublime use of vocabulary, meaning everything and nothing.

5. Stay calm. You are responsible for a rather generous spillage of acetic anhydride in a wildlife park and your questionable business practices are leaked to the media. Seize the initiative. Announce a list of voluntary schemes to clean up business practices in the future. 'We are committed to best industry working practices and are currently working on a plan to reduce leakages.' Make sure no timeframe is revealed.

6. Support human life rather than nature. Yes, you are killing the planet, but it is too expensive to avoid. Join forces with anti-slavery and human-rights groups. Sign up. 'We are committed to ending the suffering of indigenous people the world over and have been actively campaigning for a fairer society.' Whatever you do, don't mention the environment. Ever.

It's a whitewash

With businesses country-wide employing these strategies, there are signs that the public is beginning to see through such laboured goodwill. In the final quarter of 2006, the Advertising Standards Authority investigated complaints concerning forty 'green' ads. Overnight, cynicism seemed to shoot through the roof. During the first half of 2007, the authority was asked to examine three hundred green 'ads'. Even the government body, the Energy Saving Trust, has accused its paymasters of greenwash, in particular for making it difficult to understand how anyone can

GREENWASH

actually reduce their carbon-dioxide emissions. Surely that is the point? Make it sound like you are doing your bit, but in essence change nothing. Failing to grasp such an essential tenet of modern environmentalism explains why the Energy Saving Trust remains a marginal voice.

More than anyone, the trust needs to learn from one of Britain's most successful businesses. Tesco learned from research carried out by its Sustainable Consumption Initiative that it would sound good to offer club-card points to those who declined to use wasteful plastic bags. Tesco then carried on producing three billion bags a year and rewarded those with lots of points with discount vouchers, thus encouraging even more consumerism. Think big. Remember the rules. At the very least, mislead your customers. You might even make some money on the way.

WHAT'S THE DAMAGE?

* Cynicism mounts among consumers and there is backlash against companies who promote their green credentials. **Likely.**
* Major oil companies announce series of campaigns including 'save the albatross' and 'protect the greater-crested grebe'. **Certain.**
* Advertising Standards Authority records complaints against 2,000 green ads in first half of 2010. **At least.**
* Eco becomes the most popular prefix on the high street by 2012. **Credible.**
* Phrase 'sustainable development' is banned by government after poll shows only one in 300 can define it correctly. **Unlikely.**

Likelihood of greenwash triggering backlash against environmentally friendly companies: 91%

The final 31 frontier

Out with the old world, in with the new

AGENDA

* Race for space
* Spend, spend, spend
* Explore new territories
* Get outa here

Some questions are bigger than others. And somewhere, locked within the crust of the moon, scientists believe, lies the answer to one of the largest – the origins of life on earth. Interest in manned space travel has not been greater since the height of the Cold War. Nations everywhere are jostling to prove their worth by unlocking the secrets of the stars above. The second space race has begun.

This new-found fascination with distant galaxies arrives with impeccable timing. Think of the opportunity cost. Ensure that a fantastically ludicrous amount of cash and scientific brain power is committed to studying the dead disc of the moon, and any opportunity to contend with the inexorable countdown to catastrophe on this earth is blown off with the force of a launching fuselage. Hurling man into space is charmingly consumptive of cash. Billions of pounds are frittered away examining the lifeless orb above. For what? A net benefit to humankind approximating, oooh, zilch. There is always the off-chance that someone might find an odd fungus trapped in a spot of ice on a planet 150,000 miles away. Yum, yet another delicacy to import, shrink-wrap and stick on supermarket shelves – and food miles would be off the scale! And in addition to this ... no, we can't think of anything. Advocates believe that valuable lessons can be learned Out There that will help us to preserve our own planet's future. Such altruistic nonsense can be disregarded. Very soon there will be no future.

So why spend time and money on your own planet when you can find a shiny new one to play with? Meanwhile, the rest of us watch as famine, degradation and pestilence encroach across our ever-sweltering planet. As critical resources are diverted from finding solutions to thwart the ecological crisis unfolding on earth, some of our keenest minds have their heads stuck in the stars. There might be more use in shoving them up their arses.

10, 9, 8 ...

His footprints have long gone, submerged beneath a tide of lunar dust and cosmic debris. Thirty-five years have passed since man last trod on the moon. Yet, when Apollo 17 commander Eugene Cernan left the Sea of Serenity in December 1972, humanity could be excused for assuming that space colonization was just around the corner.

But that was it. The space race faltered. As the moon settled back to its 4.5 billion years of solitude, hopes faded as to when man would return. Humans, it seemed, were destined for little more than to faff about in low orbit aboard planes and the less than entirely reliable Space Shuttle. And that suited Britain just fine. In the decades since Cernan's voyage, Whitehall has rarely concealed its distrust of manned space flight. Too expensive, for starters. So much hassle for such little reward. What could chucking an astronaut in space ever truly achieve? Thank your lucky stars for the USA and NASA. They believed.

And how times have changed. Reports can be confirmed of a dramatic volte-face in the British government's reticence in hurling men into outer space. Suddenly it wants in on the colonization of the cosmos. When NASA unveiled plans for a permanent settlement on the moon by 2020, Europe couldn't contain itself. Britain swooned. Days later, Britain's then science minister Malcolm Wicks hailed interplanetary space travel as 'one of the marks of a great nation'. But while contemplating the void above, Wicks seems to have suffered his own personal brain drain.

As energy minister, the father of three had for the previous three years been privy to the most sensitive data chronicling the threat from climate change. A seemingly endless list of sensitive data told him that his grandchildren might live in a very different world. Had he forgotten? Maybe not. By promoting a space colony in the heavens above, had not the career politician dug the ultimate escape tunnel? Suddenly, the travails of pesky little earth didn't matter. Getting the hell out did. In his own words, 'For anyone to say British men and women will never explore space in the future is just a nonsense,' or, to cut the hi-fallutin' science talk, 'That's it, I'm off.' The tunnel may not be for any Tom, Dick and Harry just yet but, in theory, once world leaders have galloped off into the galaxy, you will be free to render your own planet null and void.

... 7, 6, 5 ...

Governments are preparing to give up on saving earth. Which, of course, is good news. A tangible shift in focus from the earth to the solar system offers a once-in-a-generation political symbolism that should be exploited by all those inspired by the concept of widespread environmental destruction.

It is important that major corporations and politicians echo the

THE FINAL FRONTIER

minister's message; space exploration has never been cheaper and, in fact, might provide tremendous value for money. Your air miles will rocket! In reality, ruined budgets and overblown expectations are anticipated. It is no secret that most money allocated to space exploration is enchantingly squandered by bods twiddling around with hyper-tech gadgetry in laboratories fixed firmly to the ground. And let's not ignore the fact that the worthiest scientific discoveries in space have emanated from unmanned missions.

Estimates cite the cost of creating a permanent new settlement on the moon at almost £500,000 billion. The government's latest analysis claims that cashing in on the space race is worth £7 billion, and that its most recent position is that space will become an 'increasingly important' asset to the economy. NASA has already upped its cash for space exploration to $18 billion a year. The Royal Astronomical Society wants £3 billion over the next two decades to finance moon landings. Well, we all want £3 billion over the next two decades ... But (yawn), where's the money to come from? Opportunity knocks. We look forward to a significant if opaque budgetary increase. The omens are positive for a complete waste of money and subsequent 'brain-drain' from finding environmental solutions.

Wicks' enthusiasm for space presumably earned him a one-way ticket to lunar colony number one, and Ian Pearson replaced him as energy minister. During Pearson's watch, things heated up quite considerably. Trying to drum up support for sending his chums into space, he explained that space could help us understand the 'changing climate'. Really, Pearson? Or understand how to *escape* the changing climate? He also said, 'Space technology is a vital part of our everyday life.' Oh yes? Which bit exactly? Was his lunch heated by rocket boosters from Apollo 11?

...4, 3, 2 ...

Early signs suggest that scientific spending to protect earth may already be falling. At the same time as the new-found space obsession began, cuts were in fact proposed to the UK's pioneering Meteorological Office, home to the giant supercomputers that plot the world's future weather. These offices first alerted politicians to the hazards of global warming and, for that reason alone, they deserve to shut. They never get the forecast right anyway.

And the good news just keeps rolling in. One of Britain's pioneering climate-change research centres, the Tyndall Research Centre in East Anglia, recently had its government funding almost halved. Its then head Sir John Schellnhuber, one of the most vociferous exponents of what climate change might mean to society, was forced to return to his native Germany. Schnell! Numerous projects designed to protect Britain from global warming have been abandoned. As the climate-change believers were sent packing and the government shared its lust for all things lunar, a £335 million British proposal was unveiled to study the effect of tremors on the planet's crust. Yet its focus was not on the impact of fresh tectonic shifts on earth but rather the history of 'moonquakes'. Incidentally, £335 million is seventy times the amount the government contributed to an early tsunami warning system after the deaths of 260,000 people on Boxing Day 2004.

Lift-off!

The government's real reason for all things spacey is more existential. They are simply looking beyond earth to ensure the survival of the species. Great minds agree. Professor Stephen Hawking believes it's time for a back-up, saying 'the survival of the human race is at risk as long as it is confined to a single planet'. In discussions between the UK and US concerning space, no attempt is made to disguise the real reason behind their proposed moon-based colony, from where they hope to launch a £304 billion manned journey to Mars. Their autonomous colony is a simple test of whether leading politicians can survive on another planet by 'living off the land'. As long as their best minds are focused on space rather than global warming, we wish them luck.

THE FINAL FRONTIER

As earth deteriorates, we are quietly confident that more and more money will be allocated to the colonization of space as the only option. Within a decade we anticipate no need to lobby for greater investment on interplanetary space travel. Expenditure will be driven by political necessity rather than any pretence of scientific advancement. NASA's proposals depict a fully functioning space colony by 2024. By then, the scale of earth's environmental crisis will be clear, and we'll need that cosmic colony.

WHAT'S THE DAMAGE?

* Public outcry over planned lunar colony as condition of earth deteriorates over the next two decades. **Highly likely.**
* Initial technological failures cause space projects to be abandoned prematurely. **Possible.**
* Change in governments sees space funding programmes slashed in favour of earth-bound science. **Unlikely.**
* Increased militarization of space forces nervous politicians to rethink suitability of spending billions on lunar exploration. **Not a chance.**
* Moon declared unexpectedly uninhabitable by findings of early unmanned probe. **Extremely unlikely.**

Likelihood that space travel consumes more and more of science budget: 80%

Appetite for destruction 32

Beef yourself up

AGENDA

* Consume more calories
* Join the fat club
* No pain, just gain

Imagine a world where fat people are heroes. We're not talking about those carrying the odd extra pound. No, it's the outright blobsome who will be the true idols of the future. At last you can breathe a long sigh of relief, stop holding in your tummy and throw out your control pants. Obese people are essential in any quest to f**k the planet. Celebrate by becoming one! Fat people don't need a manual to find out how to chew this planet up and spit it right back out; they just do what comes naturally.

Girth of nations

Once you've stepped off the scales for good, you must ensure that everyone else does too. Make sure that people become so fat that they can barely walk, that they morph into folk who have to drive to the bottom of the garden and order a taxi to cross the street. Encourage everyone to lie around watching television in their overheated homes. Make them feel proud to be a burden not only to the state, but to the entire planet. Pump resources into looking after them as if you were rearing geese for the finest foie gras and ensure they feel like a valued part of society. Extol the virtues of eating big, big dinners; encourage them to hunt out the most high-calorie, high-carbon, wasteful meals around.

Currently, two-thirds of UK adults and a third of children are overweight or obese. Ministers predict that, without government action, this figure could rise to almost nine in ten adults and two-thirds of children by 2050. But the drive to real portliness has yet to start. The planet's most obese country offers us a clear lesson on what is possible. The US is constantly raising the bar on what can be achieved by high car-ownership, calorie-dense fodder and cheap energy. There, the average citizen consumes 19 tonnes of carbon a year. By contrast, Britons use 10.25. Whichever way you look at it, the situation is becoming embarrassing.

Soon enough, the rakish forms of sustainable-living types will become so rare they will be looked upon as walking freakshows, and not just because they are still walking. Anorexia will be defined by any adult below a size sixteen. In your vision of the future, models will waddle along reinforced catwalks. There will be no size discrimination then; the human form will be infinitely beautiful. And the more of it, the better.

Burger me

Consider the following question: 'Are YOU man enough?' When Burger King unveiled this slogan as part of their advertising campaign, their appeal to masculine pride (not forgetting stomach-minded feminists) was brilliantly devised to promote the Double Whopper With Cheese. A delightful little offering, the product offered a neat 923 calories in one fast sitting. A typical man would need to walk 9 miles to burn off a Whopper, though once you've started gorging on this modest feast, 9 metres will seem an achievement. Sales of the burger could go through the roof. A strict diet, centred around this sublime creation, is what you must recommend to help Britain's collective waistline expand at a gratifying rate. Most of the UK population lives a reasonably short drive from one of the chain's 650 branches, so everybody can join in. No excuses please. Globally, Burger King has 11,220 outlets serving 77 million people weekly in sixty-one countries – an okay start, but one hopes that increased Whopper demand will stimulate a fresh phase of international expansion. Of course, all new branches will be drive-through outlets, as the concept of passing trade on foot becomes increasingly laughable.

Try not to settle for just the 380-gram Double Whopper With Cheese. Order large fries and a cola to ensure a combined calorie injection of around 1,500, almost the daily recommended intake for a woman and nearly two-thirds that for a man. Followers of your diet should aim to use seventy out of their ninety main meals a month in proving they are man enough. Though, be warned, we don't want you going all Morgan Spurlock on us. The original king of the burger, 37-year-old Spurlock filmed a documentary entitled *Super Size Me*, which chronicled his month of eating only at McDonald's. He was sick on the second day. By the end of the month he had gained an extra chin and grossly enlarged his liver. It's not an ideal approach. Followers of your diet must stay healthy enough to live and ideally should retain their appetite, so supplement the Whopper diet with out-of-season fruit and vegetables imported from the other hemisphere.

Meat and greet

Take a look at your beefy accomplice in its finest detail. The bap is reassuringly generous in carbohydrates while the gherkin is flown in especially from California to provide nutrition. The burger itself offers a reasonable return of 57 grams in fat, more than half the

APPETITE FOR DESTRUCTION

daily allowance, of which 13 grams are saturated and 2 grams are trans-fat, a form that is not digested normally by the body and may increase cholesterol. Don't worry. If you do get poorly from cardiovascular disease or diabetes, or suffer respiratory complaints, go straight to hospital. Southampton General Hospital is recommended, chiefly because it once conveniently installed a Burger King on its premises. There's no reason why hospitalization should come between you and your burger. Keep an eye on your children: chronic obesity means that some might die before you do and, if the planet somehow outlives the current generation, you are going to have to keep your over-sized youngsters alive to carry on the work you have so diligently started. If your fattest offspring are finding their dietary demands a struggle, drive them to the children's ward at Addenbrooke's Hospital, Cambridge, where every Saturday it is Burger King Takeout Night.

Stuff your facebook

Time to spread the word. Instigate a campaign on a social-networking site. Petition to get 'Are YOU man enough?' back on television screens. A similar campaign to save the Wispa chocolate bar worked like a dream. Be quick-minded. The government is currently considering whether to apply the 9 p.m. watershed to adverts encouraging fast foods. The Spanish government even attempted to ban Burger King's advert itself. Health campaigners complained that it promoted 'excessive consumption', which, surely, is the point. Burger King maintain that it is all about choice, which, in a capitalist democracy, seems to be fair enough.

Watching television shouldn't come at a cost to your burgeoning waistline or carbon footprint. Follow these simple steps to maximize returns:

1. Push sofa into centre of lounge (being careful to avoid over-exertion).
2. Arrange pile of remote controls by your side.
3. Turn on large plasma television.

4. Activate surround-system and seventeen micro-speakers.
5. Switch on other wide-screen televisions (hooked up to DVD player and games console). Mute them.
6. Activate central heating full-blast, ensuring you strip beforehand if it's summer. Sweating might result in you shedding weight, and that must be avoided.
7. Order a takeaway or remind a like-minded friend that it is their turn to post a BK Whopper meal through your letterbox.

To keep your mind active whilst lying on the sofa you regularly delight in mental arithmetic. You know that for every journey under 3 miles for which you don't walk, you add another couple of kilos to your carbon total. Similarly, you understand that you can quadruple weekly energy use by never turning off the central heating. You will never go near a bicycle again after recently discovering that a 20 per cent increase in cycling would reduce Britain's carbon emissions by 35,000 tonnes and turn your calf muscles into toned monstrosities. After all that exertion you will probably fancy some sleep. It's worth investing in an eye mask and earplugs to drown out the electrical appliances and lights in your home, which are eternally left on.

Absolutely burgered

Regardless of how many participate in the Whopper diet, obesity rates will assuredly grow. More sugar is being added to processed food all the time, with studies revealing that the amount has doubled in the past three decades. Appetite for self- as well as for planetary-destruction will not dissipate, despite new schemes being hatched by the government. Attempts to get a simple 'traffic light' labelling scheme introduced on to packaging of convenience foods is being admirably resisted in Brussels because of fervent lobbying by food activists including Tesco. For that, as for so many other things, you can be grateful to Britain's biggest supermarket.

Government predictions estimate that, by 2050, obesity will cost the UK a whopping £34 billion, a satisfyingly grand amount that

APPETITE FOR DESTRUCTION

will be diverted away from tackling environmental breakdown and a figure 170 times larger than the sum earmarked for developing international research on climate change. As time drags on you will find yourself increasingly up against the health police, but as Britain is already in thrall to fast food, no matter what, they'll never be able to ban the burger. Approach every supper as your last and perhaps you really *can* consume the planet.

....

WHAT'S THE DAMAGE?

* In late 2010 an inquest hears how 6-year-old twins from Manchester have died from over-eating. Coroner describes their bodies as being embalmed in fat. **Likely.**
* Burger King, whose international outlets have topped the 17,000 mark, unveils the Goddess (With Cheese) in 2013. Half a metre high, with 1,400 calories, the burger takes the world by storm. **Plausible.**
* The World Health Organization condemns fast-food chains for 'crippling' a generation. **Probable.**
* By 2020, more than a third of European adults are diabetic or have breathing problems due to weight issues. **Almost certain.**
* Mass overhaul of bus, train and plane seating dimensions announced in 2015 after majority of people can no longer sit down due to girth issues. Steel-supported beds become the norm. **Feasible.**

Likelihood of three-quarters of adults in Western countries being diagnosed overweight or obese by 2020: 89%

Flying low

Full throttle on the stratosphere

AGENDA

* Take off on a holiday or two
* Become car-free
* Don't miss out on emissions

Warsaw, Waikiki or Washington. Where to next? Hop aboard, explore the world. The further the better. You just love having your knees around your ears for eight hours with only the cremated remains of a cow in a tinfoil trough for sustenance and nothing but a film about a talking dog (which you already watched last year for some inexplicable reason) for diversion.

The appreciation of this sky-high activity will only grow. Planes might only contribute a modest amount to the world's carbon emissions at the moment but, according to UN forecasts, their contributions will treble to 15 per cent over the next two decades. Aviation is the fastest growing source of greenhouse-gas emissions. A tonne of plane pollution ejected directly into the stratosphere is almost three times as harmful as a tonne emitted at ground level.

The sky's the limit

To get more people flying you are going to need another bloody big runway. Where better than at Heathrow airport, already one of the most enduringly polluting sites on the face of the planet? This won't be easy. First, the government would have to break its historic pledge to the people of London that there will be no new runway at Heathrow. There's also that naive promise they made about pegging Heathrow flight numbers. And, of course, the government has sworn to cut carbon-dioxide emissions by 60 per cent by 2050. If you still truly believe that you can get this third runway – and therefore more flights, cheaper air travel, and the continuing incentive for hypermobility – you must accept that you will have to influence government policy. You will need unparalleled access to key figures and internal Whitehall documents. You will need your friends in Big Carbon like never before. You will need the most fearsome lobbyists on the case. With these, you might be able to mould government policy towards doomsday with all jets burning. Experts at the Tyndall Centre for Climate Change Research calculate that to achieve lower carbon-dioxide targets without reducing the role of aviation, then all other emissions – cars, industry, homes, everything – would need to hit zero. The earth can either have international aviation growth at the present rate or a stable global climate. It cannot have both. You know your preference.

Go figure

Could allowing an extra 2.5 million flights to land at Heathrow every year be achieved with no obvious increased environmental damage? On paper, it looks impossible. Only one source of hope remains. It entails playing dirty, but since when has that proved a problem? You must fiddle the figures.

The email that changed everything arrived on 9th February 2007. Sent to the Department of Trade and Industry from BAA, its arrival signalled the moment that the third runway was back up and running. Titled 're-forecast', it explained how the government could dump initial findings, which had ruled out the extra runway on environmental grounds. Instead, they could recalculate using fresh figures, mainly provided by BAA. Documents obtained under freedom of information laws reveal how senior officials from BAA were given access to Whitehall so they could cherry-pick alternative input data for their revised predictions. Figures that revealed how the expansion might cause unlawful levels of pollution and noise were sensibly downplayed. Eventually, after some admirable tinkering, they got it right. A new airport the size of Gatwick could, in fact, be bolted on to the northern perimeter of Heathrow with no adverse impact. Whoever wrote the final draft did a laudable job of ensuring it was comprehensible only to lawyers and aviation anoraks.

Friends in high places

In hindsight, it's clear you never had to worry. The government's small army of PR experts and advisers regularly flit between Whitehall and BAA. Already, a fortune has been spent on lobbying, lunches and MP upgrades. BAA funds two key groups to promote the huge social and economic benefits of airport expansion. FlyingMatters and FutureHeathrow are led by former industry and energy ministers. While the third runway consultation was ongoing, the latter held a reception in a dining room at the Commons. You remember the date vividly because you were not invited. Ruth Kelly, the current transport secretary, was, however. She spent the evening rubbing shoulders with BAA executives and other eminent characters from the proud aviation industry.

Perhaps they were looking to headhunt new political talent. The aviation industry has long used Whitehall as a recruiting ground for new personnel. Among those you know are Jo Irvin, now a member of Gordon Brown's inner circle, who once headed BAA's public-affairs department. Another paid-up Labour member is Stephen Hardwick, former director of public affairs for BAA. The lobbying firms used by British Airways – Brunswick and Lexington – also enjoy close links to Labour. And Kelly's former transport adviser Rod Eddington believed that constraining aviation would come at a 'significant cost to the economy'. His is a learned mind and his wisdom is founded upon experience. You wouldn't expect anything less from a former chief executive of BA.

Mission: emissions

With all that going on, no wonder that an airline fuel tax has yet to be imposed, a tool that would catastrophically reduce the demand for aviation. As it is, airlines are not yet subject to any levies in the form of fuel tax or VAT. You must support this lack of tax wholeheartedly, because any plans to double air travel over the next two decades are reliant on keeping flights damn cheap.

Airlines will soon be joining the EU emissions scheme and this will become the *coup de grâce*, proving to the sceptics that flying more often poses absolutely no harm to the planet. Reports will show that not one extra gram of carbon dioxide has resulted from the gigantic expansion of Heathrow. The unswervingly ingenious emissions scheme allows airlines to keep growing without doing a jot to cut back on their emissions. All they need do is buy permits with their profits and merely fly a little more often. They can even buy carbon certificates from other industries, in the hope that someone else will cut their respective emissions and thus allow aviation to reach its full potential. It is a ticket for growth. And mega-profits. Airlines stand to make billions of pounds in 'windfall profits' from the very emissions trading scheme that was designed to make them pay for environmental damage.

It is hardly surprising that Belfast, Birmingham, Bristol, Cardiff, Manchester and Stansted are all looking at major expansions to their airports. Under Labour, the UK's aviation emissions have risen from 5 million tonnes a year to more than 18 million, but the government sensibly ignores such piffling increases. The official figures calculate that targets to cut carbon emissions are on track. Dispiriting news, until you learn that the official sums do not include aviation. If they did, emissions would be 12 per cent higher and environmental targets would not even be within sight of the stadium.

Spread your wings

Brunswick and Lexington will be asked to echo the government's mantra that a third runway is essential to ensure economic progress. If it is not built in Heathrow, then other airports in Paris and Amsterdam will simply take advantage. And, of course, the third runway is far more than a 2,800-metre strip of tarmac, it is a statement of intent; that the needs of the people and their lust for cheap flights will always prevail over the environment. It is the ultimate political test of our generation and one you must win if you are to show the greenies that they can never triumph.

One day you expect Heathrow to have four runways. Seven. Eight. Ten. All of Berkshire should be designated a special landing area. Windsor Castle will become the control tower, the Queen moving out on the condition that she can have her own jewel-encrusted landing bay. Current predictions forecast that, by 2020, three times as many Britons will fly on a regular basis. Ultimately, you expect everyone to be flying weekly. In turn, rail investment should be scaled back to facilitate the needs of extra flyers. The third runway will increase Heathrow's capacity from 480,000 to 702,000 flights. You want it measured in millions.

Of course, other benefits aside from kudos will result from the additional runway – namely, the extra 25 million cars travelling along the M4 to get to the airport. They'll have to widen the roads. Possibly build some more. Relax, the government seems keen. A national road-pricing scheme to reduce traffic and greenhouse-gas emissions has been rightly jettisoned. Motorways will be widened to help road transport to contribute to the 12 per cent increase in carbon emissions already recorded under the current government.

And so the message you must send out is this: fly wherever you can and whenever you can. Bali, Beirut, Birmingham. See old friends. Meet that faceless business contact. Online is over. Forget the phone; the internet is for geeks. Take pride in being a 'people person' – everyone knows you can't beat old-fashioned human contact. Get out there. Spread the carbon.

WHAT'S THE DAMAGE?

* Number of people flying continues to climb. By 2015 two-thirds of Britons fly at least five times a year. **Likely.**
* Third runway gets go-ahead. Huge protests congregate at Heathrow perimeter fence. Tabloids dub heavy-handed police response with the headline 'Greenham Common II'. **Predictable.**
* Parliamentary inquiry into alleged collusion between BAA and the government finds 'uncomfortable' working practices, but clears both parties of wrongdoing. **Certain.**
* Flying becomes taboo. Aviation enthusiasts become increasingly viewed as the new pariah in an age of environmental enlightenment. **Possible.**
* Flying becomes the cheapest form of travel by 2014. People prefer to fly distances they would once rather have driven. Tesco unveils budget airline. **Probable.**

Likelihood of aviation remaining fastest source of greenhouse-gas emissions by 2015: 92%

Nuclear 34 wasters

Put 'em to waste

....

AGENDA

* Do some serious train-spotting
* Take the cask to task
* Cause commuter chaos

The unthinkable. A nuclear attack on Britain. An invisible swirl of radiation renders much of the home counties uninhabitable. Initial death tolls are estimated in the thousands. The emergency services are overwhelmed. All movement is banned by the government. Within hours, the nation's capital is closed for business.

No dirty bomb was smuggled into the UK. There was no remote strike from an atomic warhead. Instead, the weapon was an everyday train, a nuclear cargo that trundles three times daily across the UK. Studies reveal that a well-rehearsed attack on the routine transport of nuclear waste could spread a blanket of radiation over more than 50 square miles, killing at least eight thousand.

A new era of nuclear power plants has extended the possibility of a successful strike. Each new plant will produce approximately 40 tonnes of highly radioactive spent fuel each year, which will need to be hauled across the UK for disposal. Nuclear power has been sold by Europe's politicians as the great panacea for climate change. There is no plan B. A successful attack against the nuclear industry would, with one stroke, undermine plans to stop the planet boiling. Europe's climate-change strategy would unravel, a continent's dreams of clean energy vaporized.

Trains, planes and automobiles

The driver saw it too late. In the middle of the tracks stood a lorry. He yanked the brakes, but the train collided, screeching to rest just beyond the level crossing near Ashford, Kent. This was no ordinary train. Its cargo was a 50-tonne nuclear flask designed to carry as much radioactive waste as a small nuclear reactor. That July morning in 2002, at the height of rush hour, the flask was returning from Britain's biggest nuclear site, Sellafield, Cumbria, with its usual consignment of spent nuclear rods from nearby Dungeness A power station. Investigators must have gasped with relief. The flask had not cracked.

For those after a quick shot of environmental cataclysm, attacking these nuclear transports is a tempting target. What's more, the routes are no secret and security is reassuringly lax. In fact, a detailed timetable of nuclear transports can be found on the internet. Almost twenty nuclear trains trundle across the UK every week. Investigations indicate that the most opportune place to launch an attack is one of the six major tunnels, including Primrose Hill and Hampstead Heath underpasses in London, through which these trains regularly pass. These tunnels can create the 'high and sustained' temperatures required to cause a radiation leak. And don't forget the level crossings. Take the Kent crossing, which is unmanned, with no barrier; just some flashing lights. And with no shortage of nuke trains speeding past, Armageddon would be quite easy. It wouldn't take much nous to load a vehicle with all things flammable, check the nuclear timetable and park up. Fine particles of radioactive material would be billowing around Kent for decades. Poor old Kent. Once the Garden of England, now an atomic elephants' graveyard.

And there's always Willesden Brent freight yard in north-west London, the same yard from which the aforementioned train left at 05:52 earlier that July morning. Most environmental terrorists worth their salt will know that nuclear trains are held at Willesden sidings for several hours. With the target stationary, access is straightforward. In the summer of 2006 a newspaper reporter wandered unchallenged and without a security pass into Willesden sidings and placed a mock-bomb on no ordinary train. It had a yellow radioactive trefoil symbol plastered on its side and carried four radioactive flasks of spent uranium fuel rods. It gets better. For ten minutes the train sat unattended in broad daylight and the driver had even left the engine running. This was the tenth time the reporter had walked into Willesden sidings. By then, he had probably learned to drive a nuclear locomotive. Now that would be a scoop. If you get so lucky, the drill is simple. Climb on board the nuclear express. Release the brakes. There's a cask on the run!

Five years previously, a report by the Nuclear Waste Trains Investigative Committee of the London Assembly warned that train operators 'must improve trackside security as a matter of urgency'. Few people from the government, Health and Safety Executive or National Radiological Board deigned to show up for the committee's hearing. What little data had filtered into the public domain revealed a real problem. In 1972 just four 'incidents' were recorded involving the transport of nuclear materials. By the time of the investigation this had risen to thirty-five. No figures have been revealed since, but trends often continue.

There is also the chance that the flasks may prove vulnerable. A parliamentary report divulged that, although they can survive a 9-metre plummet, even robust Type B containers – used to move plutonium as well as spent fuel – may be susceptible to attack by missile. Written in the months before the July attacks on London, the report failed to consider the additional threat of British-based suicide bombers. Or that, one day, some pretty bright people might want to make their mark on the world using a nifty little handbook.

Police investigating the 7/7 terrorist attacks in central London unearthed a number of interesting bits and bobs. Among them, maps and photographs of the Sizewell B nuclear reactor in Suffolk, and an analysis of radioactive materials. Environmentalists have previously wandered into Sizewell B. They weren't lost and they did not enter disguised as employees. Or even terrorists. They were dressed as missiles. They had popped out in their favourite nuclear-weapon costumes. They carried suitcases with the word 'bomb' embossed on the leather casing. Yet, despite these concerns, industry secretary John Hutton announced in 2008 that Britain would be building a new generation of nuclear power stations. He hailed nuclear energy as 'safe and affordable' with absolutely no mention of terrorism or security lapses.

Within hours of Hutton giving the green light, energy giant Electricité de France (EDF) declared plans to build four of Britain's new plants. Everybody was happy. But there was another timely omission. EDF had claimed that the 'fail-safe' designs for the new plants could withstand attack by plane. Officials at the nuclear giant had calculated that 100 tonnes of aviation fuel would burn themselves out in just two minutes, causing minimal damage. But a leaked internal document showed that they had neatly failed to mention the possibility of fuel vapour forming within the reactor and exploding. They also dismissed the notion that terrorists would have sufficient skills to pilot an aircraft directly into a nuclear plant. Ironically, it would later emerge that one of the would-be suicide bombers who drove into the terminal at Glasgow airport in July 2007 had trained in aeronautical engineering in Belfast, just 70 miles across the Irish Sea from Sellafield.

Britain's nuclear plants are, in fact, extremely vulnerable to a 9/11-style attack or accident. Government's analysis confirms that older plants have 'design characteristics which make them more vulnerable to terrorist attack'. Weaknesses which cannot be divulged here. The no-fly zones around such plants are easily and often breached. According to declassified MoD documents, there were fifty-six alleged breaches by military aircraft between 2000 and 2003, one so close it activated intruder alarms at the nuclear plant's perimeter fences. Since 1999 there have been seventy-one complaints of civilian aircraft flying too close. A crash into the massive tanks of nuclear waste at Sellafield could, say experts, cause 'several million fatalities'. Logistically, it might be too complex to engineer, but the point is that the plants are assailable. It is no secret that terrorist organizations, including the IRA and al-Qaeda, have looked with interest at the colossal nuclear-waste dumps in Cumbria. In the last nine years the stockpile of plutonium has doubled. In the same period, security services have identified the existence of two thousand British-based terrorists operating in around thirty cells. More, they admit, are out there. Just 6 kilograms of plutonium was used in the bomb which flattened the Japanese city of Nagasaki. Beneath Cumbria lies enough plutonium for more than 17,000 similar bombs. A year after the London explosions, an independent panel of government experts said they were unaware of any cogent attempt by government to address the vulnerability of Sellafield.

New Labour goes nuclear

It has become clear that New Labour is very much Nuclear. Gordon Brown admits that it was a hard decision to make, but it seems he had no qualms mixing business with family. After all, his younger brother, Andrew, was EDF's head of media relations in the UK. Perhaps he turned to his best friend for advice, education secretary Ed Balls, whose father is the former chair of the Nuclear Industry Association and a director of the Nuclear Decommission Authority. The government's unswerving support has given you time to plan and execute your plot. Brown's new nuclear stations are not scheduled to be built until 2018, and not one reactor has ever been built on time. A £20 billion depository to safely secure Sellafield's waste stockpile is expected to take a quarter of a century to develop. Sites for EDF's plants have been named in Kent, Suffolk and Essex and will ensure that trains continue to pass through the capital with a potentially greater cargo of radioactivity than ever. And yet the capital's population remains in the dark, the public unaware of what is trundling through their city almost every day. So far, no fatal accident or attack has occurred, despite the trains having travelled more than 6 million miles to date. Look to the past for precedents. Once the Zeppelin was deemed safe. The *Titanic*, of course, was unsinkable.

WHAT'S THE DAMAGE?

* Nuclear transport security is vastly improved. Armed escorts are introduced. **Possible.**
* Nuclear power falls dramatically out of favour after leaked report admits the government ignored safety warnings about building plants on the cheap. **Impossible. Nukes are back for good.**
* Another accident occurs at the same Kent crossing – except that this time the train is travelling to Sellafield. **Plausible.**
* Terrorists strike train with explosive device at Hampstead Heath tunnel in 2013. Flask somehow remains intact. **Likely.**
* Another Chernobyl occurs in 2017, this time in Germany. Government retains unshakeable faith in nuclear. **Conceivable.**

Likelihood of successful nuclear attack in Britain by 2020: 17%

Going bananas 35

Packaging: it's in the bag

AGENDA

* Plastic bags a-plenty
* Protect yourself against marine life
* Bespoke packaging for your veg

Every worthwhile campaign needs a symbol. WWF has a panda, Friends of the Earth hides behind an ambiguous green circle and yours, well yours could be a plastic bag, emblazoned across the middle with the phrase 'I am a plastic bag'. An exemplary choice, you'll agree. A transparent 'thank you' to the profligate waste, packaging and food miles of the supermarkets, whose unswerving determination to ruin the planet has, at times, been quite humbling. There is no more superior icon of Europe's impulsive throwaway culture than the plastic carrier. Innumerable bags bearing your symbol will turn up on remote mountains and beaches the world over, a none-too-subtle reminder to nature that you are here and you mean business.

Well handled

Hardly Gaia's buddy at the best of times, supermarkets have been a superbly reliable ally, helping you to carry home your shopping conveniently in one of the 13 billion plastic bags handed to Britons each year. The UN claims that 20,000 pieces of plastic litter rest snugly within every square mile of the planet's surface. Come back next year and there'll undoubtedly be more. Scientists have calculated that 46,000 pieces of plastic, many of them bags, swirl around in each square mile of ocean. Environmentalists widely claim that more than a hundred thousand marine mammals are killed by plastic in the sea each year. In the future, when sharks are caught and their innards sifted for human remains, scientists will hardly raise an eyebrow to discover nothing but balled-up Asda bags. You need never be scared of the sea again. Wade in bravely armed with a couple of carriers to double up as a flotation device and a convincing anti-shark solution. *Jaws* would have been quite a different movie with some plastic bags on the scene.

Estimates suggest it can take up to a thousand years for a plastic bag to disintegrate, which means only one thing: they'll be flapping about long after you've left town. Their beauty is not only their long-lastingness but also their ability to photo-degrade, poetically breaking down into smaller toxic bits able to contaminate soils, waterways and oceans. If you succeed in trebling the pieces of plastic found in every square mile, soon every throat and gullet in the animal kingdom will know what it means to choke on a plastic supermarket bag. One day, the landscape will glimmer with plastic while, in the oceans, jellyfish will find themselves outnumbered by lifeless plastic blobs, the spitting image of themselves. What a sting.

Take the wrap

Nature, admittedly, has done a sterling job in wrapping fruit. In some ways the banana is fast food, nature's way. Encased in a protective coat, easily carried and with a simple to operate zip-and-eat design, evolution has made a pretty decent fist of it. As always, though, humanity holds the trump card. And as you wander round your local supermarket, there it is, the banana, laid out in a Styrofoam tray, a plastic film sealed snugly around its bent frame. A polythene-coated delight. You pick it up, followed by a sprout from New Zealand wrapped in its own spherical little package and, finally, an Argentinian avocado, unnaturally shiny from its plastic shrink-wrapping. At the checkout you are asked if you want a plastic bag for the three items. You nod. 'Three please.' You exit the supermarket and empty the bag's contents in the bin outside. Britain chucks away 3.3 million tonnes of food each year and here you are feeling grumpy that you've only thrown away 2 kilograms. Still, Every Little Helps.

You sashay over to the end of the crowded supermarket car park, hold all three carrier bags in the air and release. The average usage of a plastic bag is eight minutes, yours clasped consumer culture for just thirty-three seconds. They embrace their emancipation. One snags on a tree and billows like a flag, another scurries across the tarmac like urban tumbleweed. You re-enter the supermarket. Emboldened, this time you buy nine bananas-in-beds-of-plastic, thirty-three foreign sprouts and seventy-eight Argentine avocados. You ask for 120 bags and wrap each item individually, explaining politely that you are in a hurry as the duty manager apologizes for the delay in bringing your entitlement of extra bags. On average, a Briton uses 216 bags a year, but you are keen to beat that pathetic figure in less than an hour. This time you release the bags above a small brook at the rear of the supermarket. You go home radiating satisfaction. In total you have released 319 plastic bags above a pretty English market town. The smug feeling soon wears off as you realize that mild despoliation is all very well, but isn't quite enough to achieve large-scale degradation and planetary suffering. You want to make a difference to the planet.

GOING BANANAS

What a carry on

You pick up the phone and dial China. You ask for Suiping Huaqiang Plastic, the country's biggest plastic manufacturer, but it has recently closed because its government has banned ultra-thin bags. And quite right too. Who wants to be scrabbling for groceries when your defunct plastic bag splits halfway down the Portobello Road? Undeterred, you try Shenzhen Delux Arts Plastic, in Guandong Province, which makes 25,000 carriers a day, a fifth bound for Europe. You check the bags are made from polyfabric petroleum. One of the many joys associated with plastic carriers is that they are, ostensibly, made from crude oil. An estimated 12 million barrels of oil are required to make the 100 billion bags produced each year. Satisfied that you have met mandatory obligations, you place an order for 250,000 of your personalized bags. You offer £17,000, around 5 pence per bag, plus shipping costs. While waiting for them to navigate the 8,000 miles to your home, you contemplate the other uses of plastic bags. You know, pulling over one's head when David Attenborough comes on TV, tying one over a car-parking meter and writing 'broken' on it in large letters. But no, yours have a more important destiny.

When the bags arrive, you pack them into your family SUV, drive to Land's End, and with the help of your children, release them above the North Atlantic. They billow away across the ocean like a beautiful cloud of white balloons. You turn away and imagine what would happen if the ocean currents rearranged themselves to create a vortex of plastic rubbish the size of South Africa. Somewhere between California and Japan this already exists – it is known as the North Pacific Gyre and fondly termed the 'Plastic soup'. That night your dreams are laced with images of tortured whales, seals, tuna. Anything really, anything that adds to the tally of 267 marine species known to have suffered because of plastic-bag entanglement or ingestion. Those chaps at arch-enemy Greenpeace have indicated what might be possible. When their trawler, the *Esperenza*, hauled in a 1-metre net from the middle of the Atlantic, they discovered almost 700 minuscule plastic fragments, including flakes from old plastic bags, and a few nurdles, white pellets like grains of rice, used liberally by the packaging industry. In every place the experiment was repeated, a rich assortment of plastic was fished out.

Gordon Brown must have been appalled. In his first major speech as prime minister he described plastic bags as the 'most visible symbol of environmental health'. Even so, he has managed to resist the temptation of outlawing the principle of free bags for all. Britain is increasingly alone in caring for its consumers. Sweden has charged for bags for more than a decade. In Germany, shoppers have long expected to pay for them. At the time of writing, Brown has sensibly refused to endorse a bag tax to be included in the climate-change bill. Doing so would only increase government costs. In 2007, Whitehall bought 1.3 million bags purely for the purposes of promotion and for disseminating valuable marketing messages. No doubt some were also used to lug about consumables which could then be claimed back on expenses – once the politician in question had returned from their second home overlooking a bag-strewn coast, that is.

It's a wrap

Of course, there would be some vegetarian sandal-wearers who narrowed their eyes as you carefully bagged up your single shrink-wrapped sprouts. Relishing your intellectual authority, you tut loudly at the biodegradable bag they have lugged with them to the counter. What a pity that it is also bad for the environment. Another well-kept secret – shhhh – is that these, too, can ruin the planet. Ethical alternatives offered by Tesco are little better than your Chinese-reared numbers. Degradable bags are still made from plastic and so still require oil. They also end up in landfills, where the sheer volume of rubbish makes it impossible for any of them to break down and so, instead, they release that laudably potent greenhouse gas, methane. Switch to paper bags, by all means.

GOING BANANAS

They use even more energy to manufacture than their plastic pals. Fabric bags, made from hemp or canvas, may last for years but take up vital agricultural land for growing food. The overriding message should be that carrying stuff around in whatever is the decent thing to do. And don't forget that plastic bags are popular for a reason. They are good at holding a few groceries. And they are as adept at soaring about in the heavens as they are at drifting in the sea. And they don't cost a penny. Yet.

WHAT'S THE DAMAGE?

* Bag backlash grows apace in 2009. Plastic becomes the new pariah. Reports of people being spat at in the street for carrying supermarket bags. Free bags become as rare as real fur. Shoppers forced to queue in their cars right up to supermarket entrances in order to ferry food home. **Imaginable.**
* A truly biodegradable mass-consumer shopping bag is invented which is both cheap and not made from oil. **Maybe.**
* Last surviving lesser-spotted wombat found choked to death on a Morrison's bag in southern Kerala. Within days Queen's corgis are strangled by stray Dixons carriers. **Unlikely.**
* European legislation bans free bags. **Possible.**
* Shoppers buy more plastic bags than ever even after 5-pence price tag becomes mandatory. Annual numbers sold reach 220 billion by 2015. **Probable.**

Likelihood of plastic per square kilometre trebling on land and sea by 2020: 40%

Great wail of China 36

We want! We want!

....

AGENDA

* Encourage expansion in China
* Prepare for the coal rush
* Get round fortnightly collections: send your recycling east

Far, far away there is a middle kingdom, a vast beguiling land which any dedicated student of the apocalypse would be advised to embrace. It is a big country with big problems, but for your purposes it brings big opportunities. China's insatiable appetite for expansion is fed by coal-fired power stations, the belching behemoths that stand proud as peerless contributors to carbon-dioxide emissions. These gases are responsible for 80 per cent of human-generated greenhouse-gas emissions. Most derive from coal. China, it seems, is a powerhouse!

Global coal consumption has risen, and China is responsible for 90 per cent of the increase. But to those for whom gloomy warnings on climate change are never quite gloomy enough, the news only gets better. China is opening coal-fired power stations at the rate of two a week. It has plans to build another 550 on top of the 2,000 already cooking the atmosphere. In comparison, Britain has a measly eighteen. Driving this growth is growth itself. Already there is a coal rush the like of which has not been seen anywhere since the nineteenth century. By encouraging China's rampant economic expansion, you can ensure that the planet slides inexorably into climatic chaos. Praise must be extended to one of the shrewder observations of Mao. This son of a farmer may have been a lot of things: prime motivator of the Great Leap Forward; architect of the greatest famine in history; instigator of a peacetime death toll running into tens of millions but, in the canon of anti-environmentalists, Mao is a true hero. Man, he observed, must 'conquer nature and thus attain freedom from nature'.

Growth, glorious growth

Make no mistake, encouraging China to expand is one of the most powerful methods of conquering nature. By doing so, the facile lifestyles of those ethical types will be readily exposed. Are you listening, do-gooders? It doesn't matter if you cycle, recycle or plant a windmill on your roof. Events in the Orient will cancel everything out. Even if the rest of the world limits carbon-dioxide emissions to current levels for the rest of the century, projections indicate that China's growth alone will ensure global temperatures rise 6°C by the start of 2100. It will never come to that, though – the earth will be long gone. China is forecast to hurl more carbon dioxide above us during the next twenty-five years than the best efforts of Europe and America managed in a century. By most reliable estimates, China has already eclipsed America as the world's largest producer of greenhouse gases. China is the number-one threat to the planet.

Without question, your ultimate objective should be to encourage enough growth that the entire population of China enjoys the same living standards that you currently accept as given. Dismantling the planet is one thing, but keeping alive economic apartheid is really rather distasteful, a touch last-century if you like. Your goal is to emancipate the millions of Chinese enslaved in agrarian poverty. You want them driving Hummers, vacationing in Blackpool and impulse-buying electric teasmaids for brewing not Chinese tea, but leaves from Sri Lanka. Get them to drive when they can walk, fly when they can drive. The carbon footprint of the average Chinese last year was little more than a third of the average Briton. If China's 1.3 billion people consumed like the residents of the UK, the planet would need to burn an extra 5 billion tonnes of coal, double car production and increase the world's meat supplies by 80 per cent. Let the Chinese have what they deserve. In this selfish world, there are pitifully few gestures of equanimity. China, what is ours, is yours.

Coal goes gold

A sublime image to move the hardiest soul. Teetering slagheaps tower above clanking factories, everything blanketed with a smog so thick the sun is barely visible at midday. It is perhaps the furthest scene from environmental bliss to be found anywhere in the world. The city of Linfen, 300 miles west of Beijing, makes Dickensian London look like a country park. Suffocated in a spectral smog, hundreds of smelting and coking plants loom through the fug. Watching the city's 3.5 million residents wander like ghosts through the soup of industrial blitzkreig is heart-warming stuff, the holy grail of those who court the lofty ambition of planetary destruction. The World Bank agrees, awarding Linfen the coveted gold medal for best city around in terms of planetary pollution five years running.

Brits abroad

Even the worst ecological degradation can be made worse. Linfen might be an inspirational icon underscoring the side-effects of

GREAT WAIL OF CHINA

China's courageous attempt to provide teasmaids to its populace but, as always, things can deteriorate that little bit more. Thankfully, the British government is as keen to help China extract coal as you are. A delegation of British mining-equipment companies was recently dispatched to Beijing to lend its support and expertise in everything from mine exploration to coal transportation.

Analysts predict that China's growth will continue to burgeon at an eye-watering pace. China's economy has been expanding at 9 per cent a year. Linfen brags a phenomenal 12 per cent rate. In less than a decade, the Chinese economy has tripled. Growth, growth, growth. Gotta keep growing. More buildings equals more skyscrapers equals more concrete, the production of which creates 8 per cent of greenhouse gases worldwide. China's new captains of black industry – the mine bosses of Linfen – flaunt their wealth with private fleets of Rolls-Royces. China's nouveaux rich have become the world's biggest market for the venerable old lady of British car manufacturing, never famous for her green credentials. And just to make doubly sure that Rolls-Royce ramps up China's carbon-dioxide emissions, the Derby-based company recently agreed to supply a hundred aircraft engines to the country. To environmental hell in a very stylish handcart. On the all-important environmental-damage front, several other British companies have already done you proud. Among the list of polluters are Panasonic and Associated British Food and Beverages, based in Shanghai. The message couldn't be plainer if scrawled in tar on a pristine icecap. Fed up with adhering to all those tiresome green regulations back home? Pack your bags and head east. Let's be honest, nobody's going to make too much of a fuss. In Linfen they might not even see your convoy of petrochemical lorries through the miasma.

What goes around ...

The huge acreage of China holds another fine purpose for the dirty men of Europe. China can be viewed as one giant wastebin. By recycling, you could in fact be helping to accelerate climate change. Tootle around the Chinese villages that process international garbage and you'll find British crisp packets and plastic bags originating from UK supermarkets. Goods manufactured in China are being shipped to the UK and then, once used, returned to China for disposal, a deliciously carbon-dioxide-rich round-trip of 12,000 miles. All in the name of the environment. Under EU regulations, member states are not allowed to dump garbage overseas but are permitted to send sorted waste for recycling. Despite attempts by both countries to halt this perverse but perfectly formed loophole, the trade continues. China sent £12.6 billion worth of goods to the UK last year and received an estimated 1.9 million tonnes of rubbish in return. Investigations found that many of the small-scale Chinese-based recycling firms pay little attention to environmental concerns. All the time, Britain supports the recycling business, arguing that it allows for a more sustainable use of world resources. Yet little is known of the rich environmental cost of China's recycling business.

Everything's gone green

Alas, nothing lasts for ever. Worrying signs are emerging that China may have passed its nadir. A growing Chinese middle class is showing concern for green issues. Mutterings about clean energy have been voiced by its government. Dark rumours persist of China building model 'eco-cities'. Can you trust no one these days? In Linfen itself, latest data reveals that residents breathed 163 days of unhealthy air, fifteen days fewer than the previous year. A China with a responsible stewardship of the environment is a heartbreaking prospect, although unlikely when all the trends point to continuing growth and an increased courtship between her and the UK. The British government will no doubt push measures to improve coal efficiency and carbon storage to justify its trade, but you can expect most new coal power stations to ignore these completely.

China hosts a thousand environmental protests a week but effortlessly shrugs them aside. So you can cling to the hope that its authoritarian system, with no real environmental checks or balances in place, will continue to embrace disaster. And why not? Most of the world's greenhouse gases were ejected by the West. It would be immorally objectionable to deny the Chinese their chance to heat us up for a change. One day, you might even trust them to warm the planet by themselves.

WHAT'S THE DAMAGE?

* China's burgeoning middle class realizes that destroying their land is not the ideal future for their children. **Inevitable.**
* Chinese government resent being portrayed as the 'bad guys' of the planet and ratify international climate-change agreements. **Unimaginable.**
* Majority of coal-fired power stations built without new clean technology despite offers of help. **Feasible.**
* Britain continues to encourage Chinese economic resurgence as climate change spirals out of control. **Certainty.**
* As Arctic ice sheets and Himalayan glaciers melt causing untold misery, international community announces a boycott of China until it reduces reliance on coal. **Never.**

Likelihood that China keeps growing as an environmental nightmare: 87%

The 37 sex factor

Breedin' hell

AGENDA

* Give birth generously
* Condemn condoms
* Instigate irresponsible sex
* Abolish abortion

Obliterating the earth should, on occasion, be fun, and there is no more dependable means than the charms of procreation. Get involved. Woo a willing partner, agree on the merits of an unfeasibly large family, and you're off.

At the time of writing, 6,783,258,547 people are on this planet. Hold on, make that 6,783,258,653. The pace at which new faces keep surfacing is one of the more heartening certainties of modern times. When the initial death toll of the 2004 Asian tsunami was declared as 200,000, few people realized that more people had, in fact, been born that day. In the aftermath of the greatest natural disaster of modern times, the planet's population stayed broadly the same. In the moments it takes to read this chapter, another 800 mouths to feed will – fingers crossed – have materialized.

Sex it up

The United Nations agrees on the one inevitable result of this constant conveyor-belt of humanity: 'large-scale, catastrophic ecosystem collapse'. The very words are better than sex. Current breeding rates must be maintained if we are to hurtle inexorably towards unsustainable oblivion. You should be optimistic. Governments can influence many things but, let's face it; one thing they're not going to curb is people having sex. Indeed, it would be even more helpful if they were to limit contraception. Large families, after all, are to be rejoiced. Don't hold fire at two or four. Have seven. Why stop at nine when ten will do?

During your lifetime, it is predicted that a population equivalent to ten times that of Western Europe will squeeze on to the planet, with every new arrival keen to live and consume as voraciously as you or I. Ten times as many cars, a thousand times more aeroplanes and goodness knows how many extra electric waffle-makers will be required. Soon we will need half as much again in food, land and energy just to stand still. When man first landed on the moon, just three billion people lived on the planet below. Authoritative projections anticipate that, by the middle of the twenty-first century, 9.5 billion people will be groping for a living on earth. If supplies of contraceptives are not improved, the planet's population will reach thirty billion over the next three hundred years. It is a laughable prediction. The earth cannot contemplate surviving for even a fraction of that timeframe. Within twenty years she'll be wishing it was all over. The planet is there for the taking. Go forth and multiply.

Failing to plan is planning to fail

Unprotected sex has emerged as an unerringly effective instrument for advancing ecological collapse. And, as fortune would dictate, a massive shortage of condoms has emerged in the precise location where the planet's ongoing baby-boom is the most pronounced. Couples in sub-Saharan Africa are no more promiscuous than couples elsewhere, it's just that when they get down to business it has a habit of producing results. There, in the arid belt across the midriff of Africa, less than a third of couples have access to contraception. Seven billion condoms are required every year in the developing world but only 724 million condoms are currently distributed to the region from Western countries. This works out as 4.6 a year for every man. With blokes claiming to have sex 103 times on average a year (so they say), the prospects for procreation remain firmly stacked towards the positive.

According to a parliamentary inquiry, 99 per cent of future population growth will occur in the developing world. Africa's population has exploded. Hundreds of thousands of extra children are demanding the earth's resources. By mid-century, a quarter of the world's population will hail from Africa, compared to one in seven now. Until world leaders invent a way of curbing sexual appetites, the babies will just keep on coming. Ironically, in the same inquiry, condoms and birth-control pills were identified as the only realistic weapon to stem this rising tide of humanity.

Presumably keen to ensure that growth rates remain high, European countries are reticent to increase family-planning funding to the continent. Britain is among those paralysed by the moral quandary in which it finds itself. Although Britain donates a total of £60 million to fund family-planning schemes in Africa and Asia, exports of contraception masquerade under the fight against the spread of HIV and Aids. The amount has, in real terms, fallen over the last six years, but officials rule out any plans to increase funding. A welcome policy, if your primary motive is to accelerate planetary blight. Even the most modest increase to funding would yield an impressive return. Analysis reveals that every £500,000 donated might prevent 362,000 unwanted pregnancies. With the right connections, birth control comes fairly cheap. A spare £1 million is enough for governments to procure 1.1 million condoms at cost-price, or 153,800 months' worth of female contraceptive pills. It does not even bear thinking about how many babies would have been nipped in the bud had the West increased its family-planning and anti-HIV support.

Rubber banned

Pope Benedict XVI is a particularly useful asset. The role of the Roman Catholic church cannot be underestimated when his total ban on condoms ensures that around one billion believers, almost a seventh of the world's population, with 400 million in Africa alone, remain theologically opposed to lowering the earth's birth rates. Unfortunately, it must be conceded that, partly as a result, eight thousand people a day are dying from Aids and other sexually transmitted diseases, but the net population growth remains reassuringly generous nevertheless. Praise, also, to the Vatican's stance on abortion. Who knows how many unwanted babies have come into the world as a result.

No corner of the planet will escape the population explosion. In Britain, five million environmental refugees from countries such as Uganda are forecast to squeeze in among the sixty million people currently shoe-horned into our overcrowded island over the next two decades. The United Nations' eight much-trumpeted millennium development objectives to halve global poverty and environmental degradation by 2015 actually omit population as a primary goal, a canny decision that has diverted billions of pounds away from tackling the issue. Tony Blair's influential 'Commission for Africa' report, a blueprint for alleviating ecological stress on the continent, likewise glosses deftly over the full impact of the population explosion.

Promise of promiscuity

In the West, encouraging people to procreate irresponsibly is a tad trickier. A moral campaign, using slick PR-type slogans and backed by political lobbyists, might be required to affirm the traditional virtues of liberalism. Its message should be simple, advocating a return to a golden age of 'free love', an era when sex without all the paraphernalia of contraception is deemed the one true expression of intimacy. In the UK, challenging mainstream hang-ups about sex might in fact prove to be among your best strategies to hasten ecological breakdown. You should be thankful that Britain's youths are sane enough to resist the evangelism that has spawned virginity clubs among teenagers throughout the US. British teenagers are, in fact, the least likely of all to need encouraging when it comes to irresponsible sex. Even before the need for more babies has been articulated, UK youngsters have risen enthusiastically to the task. Not only are they the most sexually active in Europe, they also boast the continent's highest teenage-pregnancy rates. MPs might want to give more free condoms to teens, but getting them to use them is a different matter altogether.

Perfidious proliferation

Elsewhere, large-scale attempts to reduce population rates are dead in the water. China's one-child policy will prove politically impossible to introduce elsewhere. Even so, the country's population grows by an applaudable ten million a year. India's attempts to encourage the snip can be shrugged aside in a country whose numbers escalate every year by a figure equal to the entire population of Australia. Millions of couples have never even heard of condoms. Research reveals that condoms made to international sizes are too big for most Indian men.

Other measures to encourage more babies need to be floated by liberal political think-tanks. Duff contraceptives are one option. Scores of Chinese factories have been caught producing fake birth-control pills made from starch and glucose, a combination that

THE SEX FACTOR

provided couples with the requisite energy and an unexpected surprise nine months later. Hail South Korea, which reimburses IVF treatment to increase fertility rates. The Japanese government – whose health minister candidly refers to women as 'birth-giving machines' – has even sponsored speed-dating groups to lure lovers together. France has become Europe's most fertile country, because successive governments have encouraged women to work and have children, rather than to choose between the two.

Overall, sexual appetites show little sign of waning and there is an enduring disregard for contraception, even where it is widely available. A final word of appreciation to the millions who will continue to indulge in unprotected sex. Their courageous efforts alone sustain the daily arrival of 216,000 babies every day. Without them, the planet's future would be just that little bit brighter.

WHAT'S THE DAMAGE?

* Catholic church reverses its uncompromising attitude towards condoms. **Unlikely.**
* People, including a new generation of teenagers, stop having irresponsible sex. **Not a prayer.**
* Western states significantly increase supply of condoms to Africa and Asia as US starts supporting family-planning clinics. **Inconceivable.**
* Governments speak out over need to stabilize global population growth. **Unlikely.**
* Population expands faster than predicted, with the 7 billion barrier broken quicker than even most optimistic predictions. Condoms in Africa become more elusive than pockets of tranquillity. **Conceivable.**

Likelihood of unsustainably high population growth: 96%

Pulp friction 38

Publish and be damned

AGENDA
* Write off the future
* Print on fancy paper
* Chop down trees
* Recycle schmichael

One day you start writing a book about how to f**k the planet. Soon you realize that the actual pages are as significant as the prose, more so, in fact. With a shiver of excitement you recognize that, if canny enough, the manufacture of the book could help annihilate earth's remaining ancient forests. Books start as trees, and this one could hail from the most pristine of woodlands, a previously unspoilt wilderness butchered and churned into pulp upon which to print your enlightening words.

It would be recklessness of the highest order not to hand-pick the forest that provides your paper. The perfect place is a sanctuary for endangered species, a tranquil land undisturbed by man, a complex ecosystem that has taken centuries to evolve. You tremble with exhilaration as you hear the imaginary screams of buzz saws scything through the trees. Occasionally, lurid visions of sustainably managed woodland appear before you. Increasingly, forest-friendly pulp plagues your thoughts. But you must stay strong. Somehow, you will succeed in ravaging the planet's ancient forests with the timeless text of truth and, one day, those who survive the impending ecological holocaust will read about these bygone forests in the very books that led to their downfall.

Paper trail

And so the hunt begins for the greatest paper merchant in the world. Of course you should start with those firms fortunate enough to be sited in areas of unspoilt woodland. First stop: Finland, home to some of Europe's last untouched forests, mostly in the Gortnamoyagh region. You anticipate a quick result but, although initial investigations prove promising, several rounds of enquiries find no one with the requisite courage to play ball. You move next door to Sweden, which also, infuriatingly, has the temerity to let ancient trees remain standing. But again, nothing doing there. On to Estonia, and it's the same old story. Italy: nope. Latvia: a blank. Further east, and your journey takes you to Russia. Surely here, in the name of Stalin, there must be one paper merchant willing to chop down trees just to ensure that an obscure writer has a platform upon which to spill their half-baked ramblings. In deepest Siberia, plenty of paper companies are rumoured to be gleefully raiding the virginal boreal forests, yet the moment they hear that your book is about the environment, interest seems to wane. And so, in desperation, you head south to the rainforests of Indonesia and, more specifically, to the enormous factories of one of the world's biggest paper companies, Asia Pulp & Paper (APP).

This is hardcore

Once, backed by Barclays and NatWest, this behemoth could be unfailingly relied upon to devastate ancient forests with flair. The country's rainforests – the final refuge of the orang-utan – never stood a chance when APP was in town. However, disquieting gossip has been circulating for some time that these boys have gone soft. You whisper a prayer and make the call. Your heart sinks. The rumours are true. APP has gone clean. The world is going mad, you think. This lot had such a good thing going. They were trusted to flood Europe with ethically unsound paper, supplying, among others, a number of leading publishing houses and, by George, they delivered. But not any more. You take stock and decide to investigate several other Indonesian paper merchants whose immaculate reputations have been forged on vaporizing precious habitats. Fresh enquiries confirm your worst fears. The days when forests were illegally logged for the burblings of writers that would never sell (and a minority that did), might suddenly be in the past. Once, more than 5 million tonnes of pulp was pillaged from Indonesia, but international scrutiny has meant that no more than 1.2 million was projected to be thieved from the jungle this year. Everyone, it seems, is going clean. Or at least trying to.

Take one hatchet ...

Only Canada remains. The country that gave the literary world Margaret Atwood and Carol Shields more crucially has loads of old trees ripe for converting into books. On paper, as it were, it looks positive. You learn that almost half of Canada's boreal forests are allocated for logging, much of this designated for paper companies. Aware that you may be building yourself up again only to get hurt, you make some tentative enquiries. Bullseye.

You have high hopes for Hachette USA. This company, whose British division is also one of the UK's biggest book publishers, and owns numerous well-known publishing imprints, including

Mitchell Beazley, Hamlyn and Orion, represents almost a fifth of the market. Hachette has been known to work with a company called Abitibi-Consolidated, a wonderfully zealous logging operation with the capability to provide more than 6 million tonnes of newsprint and commercial paper a year. Abitibi-Consolidated logs trees in Ontario and Quebec, smack in the middle of Canada's boreal forests, the largest ancient woodlands of North America. This great, sickeningly intact sprawl of trees accounts for a quarter of the world's remaining forest. 47.5 billion tonnes of carbon are stored within its soils and leaves, seven times the amount produced worldwide by cars, coal and electricity every year.

But Abitibi-Consolidated is no slouch. In fact, it is pretty damn good at its job. Less than 20 per cent of the forest it has logged in Ontario remains intact. Once its diligent employees have hacked down trunks, scraps of woodchips are flogged to a company called SFK Pulp, which scrunches the remains of ancient trees into a gluey mulch and forwards the sludge to the vast paper mill of the planet's second largest producer of magazine and book paper, Stora Enso, in Germany. From there it has been flogged to lots of publishers, including your friends at Hachette. Result.

Barely able to believe this turn in fortune, you investigate further. Hachette has used woodchips from an area of degraded boreal forest a hundred times the size of London. These boreal forests were home to scores of endangered and rare species. Hachette seems like the dream partner in crime. Plans for an exhaustive, forensically researched book about pretty much nothing are drawn up. It will be very long and trivial. It will use a lot of trees. But then something tragic happens. Hachette suddenly loses its credentials. It becomes a turncoat. Overnight, Hachette becomes the enemy, their representatives suddenly talking about the importance of recycling, sustainability. They even mention the C-word. Certification.

Tasmanian devil

So the hunt must continue. You return once more to the southern hemisphere and this time to the pretty island of Tasmania, home to some of the tallest trees in the world. Such potential trophies are worth a 12,000-mile journey in – or, rather, *for* – anyone's book. The trip takes you to the door of a new £800 million wood-pulp mill built by lumber giant Gunns. Here, millions of tonnes of timber will be mulched. Some people on the island fear that the mill has the potential to pollute the Bass Strait, the slip of water between the island and mainland Australia, not to mention despoil a beautiful island. With a sense of growing agony, you hear that the government has imposed forty-eight environmental conditions on the mill. Spill so much as a droplet of chlorine dioxide and it's into the dog house for you. It was a dark day indeed when Gunns confirmed that they were going to comply with such fervent bureaucracy.

'Ancient forest friendly': egregious epithet

In mid-2008, a sad truth slowly emerges. It is fast becoming impossible to write on paper which is not corrupted by a high recycled content or approved as environmentally friendly by those chaps at the Forest Stewardship Council, who, quite frankly, are becoming the bane of everyone's life. The virtue of choice has been stolen from us all. It all suggests that your book might have to be printed on forest-friendly paper after all. If so, it is guaranteed to make for depressing reading. Maybe it will get even grimmer. Soon, more books will be published online, completely eradicating any remaining potential for chopping down attractive rainforests. But hope must spring eternal. Someone, somewhere, must still be cutting down ancient woods to ensure crappy books gather dust in backstreet bookshops. Keep looking. Tenacity is so often the principal attribute to those intent on environmental mishap. It's all too easy to just give up.

WHAT'S THE DAMAGE?

* Forest Stewardship Council is hit by corruption scandal when it emerges it has been certifying books made from Peru's last great forests. **Highly unlikely.**
* Gunns paper mill haemorrhages 20,000 gallons of chlorine dioxide in mid-2011 after being struck by sea surge blamed on climate change. It is Australia's worst ever pollution scandal. **Imaginable.**
* More books appear directly online as feckless publishers become obsessed with their carbon footprint. **Inevitable.**
* A new generation of paper merchants emerge in south-east Asia, undercutting competitors by illegally chopping down rainforest and making good old paper like in the good old days. Consumers don't give two sods. A book's a book, innit. **Possible.**
* Guide on how to f**k the planet published in late 2008. Readers complain it lacks integrity by being published on recycled paper. **Predictable.**

Likelihood of forests still being destroyed by book publishers by 2015: 43%

The butt 39 stops here

Going up in smoke

AGENDA

* Send smokers outdoors
* Dispose of butts liberally
* Piping hot on the patio

There was a time when you could sit indoors, fag drooping happily from your lips. Offices were like slow-release gas chambers where you mutually poisoned fellow careerists. You'd politely stub out your fags in an ashtray and dispose of the ends in a sensible manner. Then came the smoking ban and outside everyone trooped. Butts began appearing everywhere. Beaches, bird's nests, treetops – you name it, you stubbed your cig out on it. With the smokescreen removed, the entire world became one giant ashtray and the potential for world despoliation became clear.

Your mission, should you fellow smokers choose to accept, is to fill every corner of the planet with these faded orange beauties. Put a butt in every square metre. Fill this green and pleasant land knee-high in fag ends. Your ultimate aim is complete immobilization: the world won't be able to move for them. Motorways will be blocked by butt drifts, primary schools lost beneath collapsed butt heaps. Stop puffing your life away in the doorway of Bella Pasta, get out into the great outdoors and start tossing. Get your butts out.

Light up, breathe in, toss out

Smoking-related litter is the most ubiquitous form of rubbish in the world. Two-fifths of the world's litter is made up of fag butts, while 79 per cent of inspected sites contain at least one. (Just one? Must do better.) The good news is that butts are non-biodegradable and loiter like furry slugs for up to a very commendable twelve years. They contain toxic residue which leaches into soil; they turn clear water brown.

Forced outside, the choice of where to lay your butt is a joy to behold. Wedge into the nearest flowerbed; shove, smouldering, into a potted plant; or simply squeeze into the cracked bark of an unsuspecting tree. Saunter, ciggie in hand, to places of immense loveliness. National Parks are good, so too are Sites of Special Scientific Interest and, of course, Areas of Outstanding Natural Beauty. Failing this, any handsome riverbank, meadow, or field will suffice. Stop, exhale and jettison butt with due nonchalance. The heart-warming sight of ciggies piled high in areas of natural prettiness will inspire successive smokers to follow your lead.

Perhaps you could organize specialist smokers' coach tours, whisking chuffers into the heart of nature's last unspoilt destinations, their backpacks laden with contraband cigarettes and spare lighters. The Lake District, North York Moors, the remotest corner of Norway's arctic circle – take your pick. Would James Dean, the sexiest man ever to have balanced a tab between his lips, have trudged over to the nearest bin or would he have taken a last meaningful pull and tossed his butt into the great plains of America? The sky's the limit. Order some extra oxygen bottles and set out to climb Mount Everest. Remove your mask and light a cigarette with every other step. At the summit open a pack of tabs. Plant a fag on top of the world.

The smell of tobacco in the morning

Sit down, light up and listen. Friends in the tobacco industry are among your closest chums and you can expect their complete co-operation in any attempt to disfigure the planet. Their tactics have helped empower some of the most prominent climate-change deniers. And they're best butt-buddies with oil giants, including Exxon – hell, you can't move for these guys in some lobbying groups. Submit plans to Philip Morris and British American Tobacco, requesting they lace their tobacco with accelerants. Tests reveal that adding extra glycerine to fags makes them burn like a fuse. Whoosh. Gone. A packet of twenty may last no more than five minutes. Imagine the increased butt-litter – and the time you'd save on making a mess! Despite the fact that nicotine levels would need to be quadrupled to ensure a suitable hit, contrarily, you might urge cigarette manufacturers to examine reducing nicotine levels to 0.01 milligrams, a request that would have those killjoys in the health department firmly on your side. Reducing nicotine levels will require a slight revision of the chemical cocktail sprinkled on to tobacco. How about at the same time adding sugar, and sweeteners such as plum juice, maple syrup, and honey to make cigarettes more appealing, and to tap into that difficult children's market? Such ingredients are delightfully effective in generating acetaldehyde in the body, which increases the addictive effect of nicotine. As a combined result of less nicotine and higher addictiveness, a compulsive smoker would need seventy-eight packets of twenty per day – an environment-swamping 1,560 butts' worth – to sate their habit.

Alternatively, why not just make butts bigger? Switch round the dimensions, making the butt three times the length of the cigarette itself. Smoking would be virtually risk-free, and it would

get that meddling health minister off the manufacturers' backs. Talking of risk, those niggling anxieties that disturb the smoker's absolute relaxation – the possibility of the butt snapping off in your hand or, horror, burning down and causing slight discomfort to your fingers – would be significantly reduced.

Don't make an obvious mistake by imploring your chums in the tobacco industry to make cigarettes more toxic. Increasing levels of hydrogen cyanide, for example, a poison which has served with distinction in gas chambers for decades, may get rid of unsuspecting animals who want to smoke your leftovers, but making fags more lethal presents a dilemma: you want more not fewer smokers. You could ask for an increase in carbon monoxide, but who needs that when carbon dioxide is already the weapon chosen as principal driver of Armageddon. One poison at a time. Cigarettes that belch out global-warming gases, though, would be quite something. The tobacco industry would probably be game.

Go on. You know you want to

At the time of writing, a quarter of European adults smoke. A reasonable share of the market, but to tip the balance you must make chuffing a majority pastime. If tobacco corporations were allowed to offer incentives to entice new customers, numbers would undoubtedly rocket. For adults, free petrol vouchers and free flights. For children, a quaint series of collectable figurines, including the lion-killer, the whale-hunter and Gordon Brown.

The one drawback, as ever with fags, is price. A concerted lobbying campaign aided by climate-change deniers and connected to European think-tanks, will be required. Perhaps it's time the government allowed tobacco companies to import billions of cheap contraband tabs and sell them under the counter. Sod the taxes – surely the desecration of the environment is more important. It's perfectly feasible. In the past, British American Tobacco has been accused by MPs of large-scale cigarette-smuggling.

Mercifully, you do not have to worry about the number of

smokers diminishing. Cigarette manufacturers are committed to reinforcing their 1.5 billion strong army of smokers. They have had to put up with all sorts of accusations. British American Tobacco and Philip Morris have fiercely denied allegations that their advertising targeted young and underage smokers in Africa in order to increase smoking rates in developing countries. They also deny adding sugar to cigarettes to woo US adolescents. Since they don't endorse these methods, a new technique is required. It will no doubt be backed by one of the most sophisticated influential lobbying groups around. Few companies are closer to the centre of government than those guys at BAT. If this lot cannot bend the rules, then no one can. Naturally, the government may choose not to listen; it may endorse a repeat of its anti-litter Watch Your Butt campaign. If they dare try it, you can show them yours.

A barmy evening

Europe's desperation to force smokers into the great outdoors boasts other plus points. Outside your local, the Chelsea Tractor, the pub owners are attempting to fry the planet single-handedly. More than 1.2 million patio heaters were ordered by pubs and restaurants in the UK in the wake of the smoking ban, doubling the existing number. Each one, used almost nightly outside its respective pub, rapturously releases as much carbon as a non-smoker flying return to New York. The government estimates that another 282,000 tonnes a year will be sent to the heavens from patio-heater-related emissions. Sadly, this offshoot of the smoking ban is to be cut down. The European parliament recently voted to phase out patio heaters.

Just fine

The government also wants to persecute smokers with £80 on-the-spot fines for people who drop butts. There's an easy way to avoid them. Buy a can of cola, a packet of crisps and a pizza. Consume all three. Smoke your cigarette and hide the butt in the empty can.

THE BUTT STOPS HERE

Hurl into a nearby stream. Place another in the crisp packet and drop in a Royal Park. Smoke one more, shove inside pizza box and frisbee into a neighbour's garden. Relax. Few have been fined.

Another idea, presumably not dreamed up in an inspiring miasma of cigarette smoke, is to make smokers apply for a £10 smoking licence. Suggest that it would make better sense to charge non-smokers. The heavier you smoke, the less you pay. A series of supplementary questions could ask what the smoker does for the environment? If they refuse to recycle or to save electricity and they own a 4x4 they only use for the school run, they will receive a £10 fag voucher each week. In your (pipe) dreams ...

WHAT'S THE DAMAGE?

* Encouraged by the warming climate, more and more take up smoking. **Remote possibility.**
* British American Tobacco begins lacing product with pheromones, nostril-flaring perfume and vitamin B6. Within weeks, smoking has never been more popular. **Imaginable.**
* Kate Moss is spotted on the summit of Mount Everest smoking a Gauloise. She looks fabulous. Smoking has never seemed cooler. **Improbable.**
* Cabinet minister caught dropping cigarette just days after 'Watch Your Butt' campaign relaunch. That night, he jokingly moons before assembled media. **Feasible.**
* Environmentally friendly cigarettes are launched in 2010. They have biodegradable butts and contain no harmful chemicals. **Never.**

Likelihood of cigarette butts being found in practically every square metre of planet by 2015: 61%

Warm 40 front

A load of hot air

AGENDA

* Don't believe everything you're told
* Play Devil's advocate on climate change
* Drum up sceptical support

Here's some food for thought: man-made climate change is a myth spun by charlatans and greenies, a load of baloney, data distorted by lobbyists for ideological purposes. In truth, the science is uncertain. There is no consensus to suggest that humans are to blame for the change in climate. Peer back into history and you'll see that the weather has always been changing. And yet it is you getting the blame. Governments aren't about to complain; the politics of fear is one of the most effective tools they possess.

Sir Nicholas Stern, the minister who was responsible for conducting a review of the economic cost of climate change, estimated that the cost of capping carbon emissions was £3.68 trillion. Absurd, when carbon emissions may not even be to blame for global warming. And even if they are, the real effects are decades away. In the meantime, everybody needs to take a reality check. Remove impossible climate-change targets and regulations for business and stop panicking. Forget the con that is climate change and choose life.

Bedfellows

Saunter past the Royal Opera House and turn right towards the buskers and tourists of London's Covent Garden. Proceed along the cobbled piazza. Stop outside the Diesel clothing store and look up. There, on the third floor of grey-stoned Bedford Chambers, is the team who will guide the next stage of your quest to cripple this already enfeebled planet. Inside are people who propagate the following: that man-made climate change is a load of bunkum. The little-known International Policy Network is a good place to hang out. This global think-tank's aim is to aid 'empowerment, respect, prosperity, health' and, goodness knows, some people could do with that. The network is an ardent voice against attempts to lower carbon emissions. Hostile to global-warming regulations, particularly the Kyoto protocol, the major agreement in place to reduce carbon emissions, the INP believes that there is no real harm in allowing carbon emissions to increase and observes that the only real achievement of taxes and regulations is to 'hinder technological innovation and economic growth'.

The gang at Bedford Chambers foresees a future of making money without all this tiresome whingeing about 'carbon footprints' and the like. Why not invite Julian Morris, the network's executive director, out for a bite to eat? There is plenty of choice in the nearby piazza, though naturally you should choose somewhere which specializes in unsustainable shrimps from Indonesia, illegal cod from the North Sea, or cheap meat from the Amazon rainforest. Make sure you impress Morris. He'll no doubt be watching to make sure you care less about conservation than he does. As for the conversation, expect it to be lively. Morris is a man of forthright views. He once described Britain's chief scientist, Sir David King, as 'an embarrassment to himself and to his country'. King's crime? He said that climate change presented a bigger threat to the world than that posed by terrorism. You might recognize Morris. He used to pop up on the BBC pretty often, courtesy of the British licence-fee payer. Fortunately, he's not one to let his mouth run away with him; his organization allegedly received payments from ExxonMobil, the oil giant which has aggressively lobbied against climate change, but Morris never mentioned this.

Morris comes across as an articulate economist, someone talking sense amid the hysterical predictions of angst-ridden futurologists. His organization published a book revealing that the world has until 2035 before it is required to take action over climate change. His argument offers a window of opportunity far greater than that which is required. By the time 2035 comes, the third-floor of Bedford Chambers might well be under water, or weathered beyond recognition by an unrecognizable heat. Obliterated, perhaps, in a nuclear catastrophe. The flurry of buskers skulking outside his window may finally have been persuaded to move on.

Go tell it on the mountain

It's time to start disseminating the gems that science has tried to keep under wraps. Brush aside the findings of the world's most eminent scientific institutions. Dismiss the thousands of papers published in scientific journals which link man to climate change. They are the conspirators, you merely speaking up for the truth.

See to it that a paper is drawn up, provisionally titled: 'Climate change: A load of hot air'. Among its persuasive contents will be the fact that more British people die of cold than from heat-related causes. It seems global warming might in fact save lives. From Bedford Chambers, this vital message will spread across the world. You could try handing over hard cash to journalists who write

WARM FRONT

articles promoting your line – £5,000 ought to do the trick, although it's unlikely you'll need to use such methods. Instead you can employ a sophisticated network of like-minded think-tanks, reaching thousands of media outlets across the world.

Contact the network's old address in Washington DC. Number 1001 Connecticut Avenue also happens to be home to one of the world's most famous think-tanks, the Competitive Enterprise Institute (CEI), who have worked commendably hard heaping doubt on scientific global-warming consensus. In a 2006 television advert, the CEI spoke out to millions of US viewers: 'Carbon dioxide. They call it pollution. We call it life.' Quite. The environment lobby, desperate to demonize the world we live in, seems to have forgotten that without carbon dioxide the world is dead already. Kaput. Gone. Humans even breathe out the stuff. And they want to reduce levels of one of the most natural substances around. That, surely, is bonkers.

The man you need to find is Roger Bate, fellow of the CEI but also a director of Morris's think-tank. Bate and Morris are old muckers and have much in common. Bate, too, has publicly derided cutting carbon emissions as 'folly'. Bate is an expert at playing the US lobbying system and his connections will be vital in spreading the anti-climate-change message stateside. Handily, Bate is also a fellow at the American Enterprise Institute, the body that encouraged scientists and economists in Europe and the US to submit articles that could help undermine a report by the United Nations Intergovernmental Panel on climate change, one of the most critical ever published on the issue.

Exploit the CEI's connections with the George C. Marshall Institute, whose chairman Frederick Seitz sensibly argues that more carbon dioxide would make the world a better place. In turn, they will turn you on to the Heartland Institute, which correctly insists climate change is the product of 'junk science'. Climate-change strategies 'will surely make us poorer'. Absolutely. If you want to make money, there's no point in faffing about with that dreadful emissions-reduction malarkey. Get that carbon dioxide out where it belongs. Your network of experienced, sophisticated institutions will disseminate the message far and wide. It may have come from the offices of Morris, but by the time it's out there, this vast network of eminent-sounding institutions will have created the impression that doubt about climate change is widespread.

And the bonds among these groups do not end there. All have received money from the world's largest and most profitable company, oil giant ExxonMobil. Morris's International Policy Network has received almost £200,000 from Exxon. The Competitive Enterprise Institute at least £1 million. The American Enterprise Institute a cool £800,000. The George C. Marshall Institute another £350,000, with the Heartland Institute receiving the slightly higher sum of £400,000. Denying climate change is pretty lucrative work if you can get it. In 2006, Exxon spent more than £1 million on forty-one leading lights of the sophisticated climate-sceptic industry. Some of them sound mightily impressive, like the Centre for the Study of Carbon Dioxide, but for some reason their work never appears in peer-reviewed journals.

Weather the storm

For almost a decade, during which crucial international talks should have been taking place to tackle global warming, the debate was beautifully stalled. They were the glory years, a period when the planet warmed nicely and a spate of natural disasters served as a warning that wasn't heeded. Now, the penny is finally dropping and you will need to hone your efforts. There are, of course, many groups who receive funding from interests trying to educate the public that climate change is not such a worry after all. A number of EU-focused think-thanks, many based in the EU capital Brussels, are suspected of receiving funding money from ExxonMobil. Sceptical think-tanks active in Brussels are, quite rightly, unwilling to voluntarily disclose their funding sources. Exxon, meanwhile, has stated that it is withdrawing funding from

WARM FRONT

climate-sceptic groups, though details are suitably opaque. No matter. Should things get desperate, you will always have David Bellamy. A respected conservationist, he is one of the few with the courage and credentials to speak out. He believes that 'no facts link the concentration of atmospheric carbon dioxide with imminent catastrophic global warming'. Bellamy is famous for his opinion pieces, which stud the world's media and reach audiences of millions. Some, surely, must listen. He might not strictly be called a scientist; rather a naturalist, but in these carbon-dioxide-heavy times, speaking up for the truth is what really matters.

WHAT'S THE DAMAGE?

* New regulations force lobby groups to declare where their funding originates from. **Probable.**
* Exxon stops support for all think-tanks and begins publicizing entire accounts in show of unprecedented transparency. **Remote.**
* International business community weighs up cost of tackling climate change against increasing evidence that efforts don't make a jot of difference. In 2012 they give up on the whole damn thing. A new era of unfettered trade sees carbon emissions rise by 16 per cent a year. **Feasible.**
* David Bellamy is knighted for services to science. **Preposterous.**
* Millions are too petrified to leave their homes for fear of adding to their carbon footprint. New network of climate-change deniers, funded by big guns of retail, travel and hospitality, arrive on the scene to coax them out. **Plausible.**

Likelihood of climate-change deniers increasing influence: 21%

Emission impossible

Guilt-edged security

AGENDA

* Go carbon neutral
* Take flight
* Offset your guilt

With everything going green – businesses, politicians, so-called friends – you begin to despair at the state of your nation. It's time to ratchet up the devious; a new form of Machiavellian manipulation is required, something that will allow everyone to carry on polluting while at the same time conning themselves into believing they are saving the planet. A scheme that will assuage the anxieties of the masses without remotely effecting change. Green taxes? Too boring. Eco-charities? So last century. Energy efficiency? No way – it might make a difference. You need something so crafty it will legitimize an environmentally destructive lifestyle, giving the country a guilt-free pass to carry on as normal.

Just when you're beginning to feel as low as the water levels about a South Indian isle, breaking news sends a ripple of excitement through your furry veins. Carbon offsetting is the answer. Scrutinizing the fine print of this very dull-sounding EU trading-emissions scheme, you feel a glow of salvation. You phone up your pals in big business – bankers, insurance brokers, trusted hedge funders and diplomats. All of them are offsetting like crazy. They can't believe their luck either. Some are even preparing their own carbon-offset schemes. 'There's money in guilt and money in fear,' they say. You promise on behalf of your upstanding, eco-aware chums to spread the word. After all, why bother reducing emissions when you can pay a couple of quid for someone in Africa to do it for you?

Would you credit it?

It's only March, but you've already done Miami, Mauritius, and the Maldives this year (and why not?), each time pulling up to Heathrow in your Hummer H3, the most adorable US carbon monster to arrive in Europe yet. In the course of your travels you've hurled several score tonnes of global-warming gases into the air above. Except, hang on, you've actually done no such thing. Yep, you've perpetrated absolutely no harm to the planet whatsoever; the warming effects of your jet-powered jaunts have done bugger all to hurt the world. You bought some 'carbon credits', you see. And this sly little purchase mitigates the tonnes of carbon emitted by your irresistible gallivantings.

Carbon offsetting is beautiful. You can do anything. And it's awfully trendy. Live like the rampant ecocide-believer you undoubtedly have become, poison the planet as far as your wallet can stretch and – the best bit – by doing so you will be lauded by your nemeses, the grisly greens. Carbon offsetting is one of the most consummate inventions of your lifetime, the sumptuous damage created by your excursions simply negated by chucking some cash at an offset firm that claims to cut carbon production elsewhere. Fly to Nicaragua, but plant a shrub in Africa. Drive to Cannes in your cherished wheels and seed a spindly bush in Mauritania that you pray will die a sapling. In truth, some poor blighter on slave-labour wages grows these weeds for you, but don't worry about that. In fact, don't worry about anything: your guilt is expunged.

In many ways, these pathetic, faraway shrubs are the new equivalent of buying indulgences, the medieval Roman Catholic practice whereby pious sinners paid priests to offset purgatory. Alternatively, if trees don't turn you on, you could plump for energy-saving initiatives, usually in the developing world. Fly to Fiji and, in return, a Bangladeshi village is given a couple of low-wattage lightbulbs or a two-pence-sized slice of solar panel. Everyone is doing it. How else do you think the 2012 Olympic Games in London could be labelled 'carbon neutral'? The biggest, baddest sporting event in modern history is not even close to clean, so the organizers have arranged for some distant trees to be planted in planetary penance.

In a nutshell, that's about it. UK Plc continues with business as usual. Having chopped down forests and whizzed through fossil-fuel reserves like there's no tomorrow, you are officially allowed to count some trees on the Côte d'Ivoire as an offset to your pollution. And it is only right that these trees are planted there. You need no reminding that the forests exploited to heat the planet are located in the countries that have contributed least to global warming.

Carbon come-on

For carbon offsetting to take off as steeply as one of your holiday jets, certain facts must remain hidden from the public. Were people to learn the truth, the embryonic carbon-offsetting revolution might falter before it has fulfilled its damaging potential. First, there is no certainty about the best way to offset. A tree will naturally soak up carbon dioxide from the atmosphere as it grows, but what if that tree is burned, deliberately poisoned, or suffers a mysterious 'natural' death? All that carbon is ejected straight back where it came from. Proponents of the scheme require these trees to grow for years – up to a century, in fact. You are buying into a 'guarantee' that your trees will not only be planted but will survive for a hundred years; only then can your emissions from the long weekend at Waikiki Beach be truly 'carbon neutral'. Many of the countries that host these projects have not even been around for a hundred years. The Côte d'Ivoire has existed in its present state less than half that.

Increasing pressure on land resources should guarantee that most trees never come close to reaching maturity. You have a list of carbon-offsetting plantations in Africa, including high-profile reforesting projects in Kibale, the mountainous region in western Uganda. You suspect that growers there will accept modest payment to 'accidentally' obliterate the guilt sponges of rich holidaymakers. Tell them they can use the land for building homes, perhaps even for growing crops for their families.

And do not forget the time-lag factor. You take the H3 on a spin from London to Newcastle, spewing forth several dozen kilos of climate-enhancing gases en route. Within three hours you glimpse the Tyne Bridge. But trees, you remember, take a teensy bit more than three hours to grow. By the time the earth starts melting, you don't expect them to be any bigger than a withered sapling.

There is also confusion over the amount of carbon that needs to be offset. There is no universal, all-seeing green-eye who calculates the cost of lavish lifestyles. Carbon-offsetting schemes do not belong to a regulated system; there are no legal standards. Working out the precise, or even vague, impact of travel is fraught. To arrive at an accurate cost of your flight's carbon trail you must know the aircraft type, prevailing winds, the route, and the number of empty seats on board. Is everyone just guessing? One offset firm claims that the environmental debt from a London to Bangkok return flight is 2.78 tonnes of carbon, which will cost a bargain £20.84 to settle. Another firm calculates it at 30 per cent less carbon but charges 20 per cent more. Others estimate the damage as reaching almost 7 tonnes. You can also do a carbon-share. Persuade several friends to buy into the same scheme and reduce your carbon-offset outlay costs. Several individuals' guilt will be expurgated, but at the cost of only one.

Come off it!

Sadly, you're not the only person to spot these flaws; some have rather frustratingly started to poke fun at this well-intended scheme. One website offers unfaithful partners the chance to become 'cheat neutral' by allowing them to offset their infidelity by funding certified 'monogamy-boosting' offset projects. The website creators share your knowledge that carbon offsetting is, at best, a sticking plaster, but you hope the public will ignore the tired satire behind cheatneutral.com, who believe that paying £30 every time you head off to the tropics is the equivalent of putting a biodegradable paper bag over your head. Keep the blinkers on. You need to silence such puerile troublemakers, and convince everyone just to ask for a shrub in Africa and get on with their life.

Big business intends to get on with theirs. In fact, the biggest polluters have already got their own system worked out. They have the EU trading-emissions scheme, designed to give businesses incentives to cut greenhouse-gas emissions at the lowest cost. Again, it is deliciously straightforward. A country caps emissions at a certain level and grants firms allowances to emit carbon up to this benchmark. Firms that emit more than permitted levels must reduce emissions or choose, as they invariably will, to buy permits from companies that have polluted below their allowance.

Experts have their doubts about governments relying on carbon trading to meet their emissions targets. After the first two years of the scheme, there was no reduction in industrial carbon-dioxide emissions, then they actually started to rise. The price of carbon was placed too low; political mathematicians had shrewdly devalued the planet. Polluters simply bought allowances on the

EMISSION IMPOSSIBLE

cheap and carried on. Business as usual. The UK has joined others in a wonderfully orchestrated carbon-trading scheme that protects protected forests in distant lands. Norway, on the other hand, announced it would inject £1.3 billion into forest conservation, more than eight hundred times what the UK has pledged. To your mind, it was far too sensible a gesture and, worse still, ran the danger that it might even make a difference. Nope, choose carbon offsetting, the only true way to see off the planet.

WHAT'S THE DAMAGE?

* Major scandal as carbon-offsetting firm admits it forgot to plant trees in Chad as promised. **Laudable.**
* Government makes carbon offsetting compulsory for all flights and lengthy car journeys. Freed from hassle by greenies, politicians dutifully begin booking seven foreign holidays a year. **Unlikely.**
* EU carbon-emissions-trading scheme under attack in 2009 as emissions from major polluters rise. Again. **Probable.**
* UK government announces massive investment in reforestation projects coupled with clean-energy programmes through increased taxes. **Not likely.**
* Every major global event promotes itself as 'carbon neutral'. No one has a clue where and how the offset is calculated. **Certain.**

Likelihood of carbon offsetting making a difference to the planet by 2015: 3%

Oh my green god

The day of reckoning is upon us

AGENDA

* Preach to the converted
* Crucify creation
* Pray for the end

> *'Be fruitful and multiply and fill the Earth and subdue it; have dominion over the fish in the sea, over the birds of the air and over every living thing.'*
>
> Genesis 1:28

And there you have it. The Lord orders Adam and Eve, frankly, to let rip. The world, the message is clear, is theirs for the taking. They are the masters of their universe. Now, whether He intended the latter-day Adams and Eves to vanquish the dodo, concrete over ancient peat bogs, scythe down the Amazon, and generally run amok is a question of interpretation. But your interpretation says very much yes.

In a way that politicians can only dream of, religion exerts the power to fundamentally change people's behaviour. Followings of billions mean that believers are the largest massed constituency on earth. Some spoilsports moan about religion being the root of all problems. Nonsense. Religion is a darn good thing and, for your purposes, well, you could hardly ask for more. Your view might, of course, change if religious leaders were to start urging believers to adopt a radical, eco-friendly way of being. But for now – and pray let it continue – they have failed to put the environment's protection at the core of their moral teachings.

Almost miraculously, religious leaders have remained practically silent or inconsistent on arguably the biggest issue in mankind's history. True, there are a few readings here and there but, largely, they have kept mum on species extinction, ecological degradation, and the general good work you've been presiding over for some time now. If you are to keep on track, you need religion, if only to help justify humanity's rightful dominion over all things winged, four-legged, or coated in bark. If the planet reaches the stage where the world's population worships the earth as God, then you can say amen to any pretence of wholesale destruction. Religion needs to keep underlining human's rightful place in the planet's pecking order – at the top, looking down on the land he has made his own, a land where Man himself plays God. And plays hard.

Knocking on heaven's door

Annoyingly, the US evangelical Christian movement has emerged to warn of the need for action to curb climate change. But, actually, this isn't a bad thing. You had hoped that, if any religious movement was going to wade into the debate, it would be this lot. Most people you know cannot take these guys seriously. Instinctively, most do the opposite of whatever they say. Even honourable member US president George Bush feels they may have pushed the boat too far this time, and has opted to ignore their demands to introduce a mandatory limit on fossil-fuel emissions.

For millions there can be no greater turn-off for reducing climate change than siding with such minds. This gang, you hope, could help to further derail the integrity of the entire environmental lobby. History shows that, once they pronounce something as good or bad, they'll push it as far as possible. Hence, environmentalism is labelled 'creation care'. What else can we expect? Lynching for homeowners who have sub-standard loft-insulation? The electric chair for anyone who grills rather than toasts their daily bread? The stocks for anyone attempting to turn their (bottled) water into wine? Already they have launched a natty little campaign: 'What would Jesus drive?', which strikes you as rather a dumb question. Any chap able to walk on water can pretty much travel as he wants.

But, tempting as the promotion of these lot is, your job is to spread the word of an even more useful wing of America's evangelists. With consummate timing, a fortunate schism has appeared within the right-wing arm of evangelism. On the one hand you have those pushing to halt climate change, and on the other those who are desperate to damn well ramp up global warming as quickly and as furiously as possible. If it's a case of choosing teams, you know who you're batting for. Your new friends are called the 'End-timers'. After careful deliberation, they have adopted the non-debatable truth that climate change is part of the prophetic prediction of the gospel. Yep, they argue that believers should burn as much fossil fuel as possible in order to hasten the second coming of Christ. By caning the planet you create instant salvation. Simple. They celebrated the Asian tsunami and Hurricane Katrina with the zeal of those on the fast escalator to 'im Upstairs. Perhaps, you wonder, their anti-abortion credentials might stem from the knowledge that, the more people consuming the planet, the quicker they'll arrive at the gates of heaven. You must make contact with these people; make them your disciples. You doubt that anyone will work harder to f**k the planet than this little gathering.

OH MY GREEN GOD

The strait gate gets broader

Closer to home, things are more measured. The Archbishop of Canterbury, head of the Church of England, faced with the weight of scientific evidence that climate change is advancing, did what any clear-headed religious head would do in the face of such a monumental challenge to the natural order. He ordered a green audit of his parishes forthwith. Not content with such draconian gestures, he released 'How Many Light Bulbs Does it Take to Change a Christian?' (£4.99 from all good Christian bookshops). There is no punchline. This, after all, is no time for idle frippery. And give or take a few well-intentioned earnest gestures, that's about your lot. Rowan Williams may have warned in his Christmas 2007 speech that the planet is not a warehouse for your greed, but one is tempted to wonder who is really listening.

Some of his colleagues can appear a little more inspiring. True, the Bishop of London blotted his copybook by condemning flying as a sin, but then he atoned by abandoning his flock during the most important week in the Christian calendar to take his wife on a paid-for ocean-liner cruise worth £7,000. One hopes that Richard Chartres actually knew that large cruise ships emit almost double the carbon-dioxide emissions per passenger per mile than a long-haul flight. Maybe he didn't care.

And so on to Catholicism, which, frankly, has umm-ed and ah-ed about the issues of environmentalism, and whose leader spent a couple of gratifying years in the post of Pope before deigning to utter the words 'climate change'. Pope Benedict XVI's predecessor, though, was more your kind of person. With the social mores of a billion followers under his influence, he simply did the right thing by electing to avoid the issue altogether – although it must be said that Pope John Paul II failed to sidestep the strictures of tokenism and did end up installing low-energy light bulbs in the Vatican. Even so, you feel confident that the religion of Catholicism remains in fine shape for your purposes. Perhaps he just wanted to avoid fading the photo-sensitive frescos.

And when Vatican officials were asked in 2006 whether they had invested in any coal or oil companies, instead of looking shocked and spluttering a forthright 'How dare you?', no one seemed to have a clue.

Islam seems even more of a boon. Few argue that environmental concerns have yet to progress beyond infancy. For starters, it's quite tricky to reconcile the return air flight to Mecca with followers' carbon footprints, and impossible to quell the pervading belief that environmental destruction is a by-product of the rich Christian West. For the majority of Muslims, Islam is more than a cultural or political identity. Religion matters, a lot. Yet look closely and green themes emerge from within the Quran. For a start, it mentions that 'the sun and the moon follow courses precisely reckoned and the stars and the trees bow themselves in adoration and the heavens, God has raised them up, and set a balance. Transgress not in the balance.' To transgress this balance is to commit a crime against God. You are concerned that Islamic teachings may one day utilize such phraseology.

You also fear Hinduism and the fact that some of its gods are part-animal, such as Ganesh, the elephant-headed deity. Yet you note with satisfaction that, in India, the Asian elephant is on its uppers and that the sub-continent seems to be riddled with an incredibly healthy fatalism as far as Armageddon is concerned. India is due to be one of the great polluters this century and has dedicated just 0.2 per cent of its budget to forest and wildlife conservation. Such a philosophical outlook can be critical in achieving environmental breakdown. In a bonkers, misfiring planet, people will eventually have to accept that what happens is God's will after all. 'Chill out – the big man knows what he's doing,' they will say when the sky comes crashing down. As your plan takes shape, people must be sanguine enough to accept the inevitable catastrophe with a shrug of the shoulders.

And so that leaves the newest faith of all and the one you most hope will succumb to the altar of hypocrisy. Its converts are

OH MY GREEN GOD

everywhere: you can spot them loitering beside the organic avocados at the supermarket, fiddling with their Fair Trade cardies. They belong to the new C of E, the Church of Environmentalism. Their recycled lives are governed by the concept of energy efficiency and good turns. While you plan to bring pestilence, war, famine, and degradation to the planet, death by boredom appears to be their chosen means of departure.

WHAT'S THE DAMAGE?

* Archbishop of Canterbury rails against the 'wasteful' West in Christmas 2008 sermon. **Certainty.**
* Extremist Islamic preachers blame the EU and the US for using climate change against them in the holy war. **Likely.**
* The Popemobile is converted into a six-cylinder drag racer using biofuels. **Unimaginable.**
* Evangelical 'End-timers' caught on tape celebrating a killer tornado across the mid-west. **Plausible.**
* Archbishop of Canterbury denounces the 'greed and moral vacuum' of the West in Christmas 2009 speech. **Certainty.**

Likelihood of the environment becoming a central tenet of religious teaching by 2015: 42%

43 Cold comfort

Rain on their parade

AGENDA

* Cool reception for planet-cooling concepts
* Cloud the issue
* Reflect and absorb

The very best scientists have been cooking up quixotic ways to undo your sterling efforts to cook the planet. Giant space mirrors, flotillas of artificial cloud-makers and 'man-made volcanoes' are all touted as a panacea for runaway global warming. You must do all that is humanly possible to ensure world governments believe their technological fix is the solution. It was technology, after all, that got us into this fix in the first place. An even bigger dose of the same might be the last thing the planet needs. Any chef who has added too much pepper then tossed in extra chilli to negate the damage understands that trying to undo mistakes can only exacerbate the problem.

Head in the clouds

You must ensure faith in bizarre and ambitious schemes continues. Doing so will ensure billions of pounds and scientific resources are diverted away from genuine attempts to curb emissions. Political will to enforce changes in carbon-rich lifestyles will fade, international protocols will be suspended, and clean technologies put on hold. In the heightened scramble to save the earth, renewable energy will be seen as a little too dull. Why stick a solar panel or a windmill on the garage roof when you can invest trillions on hurling tiny mirrors into outer space?

But the real bonus in these schemes lies in their ability to let the world carry on with the status quo. While hope of a quick-fix solution remains, humanity will continue pumping 8.5 billion tonnes of carbon dioxide into the atmosphere, to levels unmatched since dinosaurs ruled the earth. Climate scientists insist that the only way to save the planet is to coax the world down to practically zero emissions over the next two decades. In essence, the promise of a technological silver bullet to climate change is the greatest smokescreen ever, the ideal camouflage for ensuring nothing ever changes. Let world leaders make their empty promises, let their people hope. Let them twiddle their thumbs as the earth burns.

It's all done with mirrors

Commendably, the sun is fulfilling her primary duties, giving life to earth. Yet, recently, the feeling is that perhaps she has gone a touch overboard. It is 2014 and, amid another hotter than usual afternoon, an international climate-change summit enters its closing stages, with leaders discussing how precisely they intend to reduce the sun's radiative influence on proceedings. Odds appear stacked in favour of controversial plans to fire mammoth mirrors into orbit in order to help block out the heat and hence reduce global warming. On hearing the news, you call up the University of Arizona and ask for astronomical-optics expert Roger Angel. You warmly congratulate him on his well-received plan to create a 1.8 million-square-mile solar shield consisting of a trillion mirrors. It will be hurled into space using electromagnetic coil guns. The cost is satisfyingly significant; even in 2008 it stood at £2.5 trillion, more than twice the entire GDP of Britain, and both inflation and development costs have almost doubled those forecasts in the time since. The vast majority of key global-research projects to tackle climate change have been shelved in order to pool funding for Angel's mirror scheme. Estimates put the reduction in solar radiation, if trials are successful, at 3 per cent, which would be enough to lower temperatures to reasonable levels.

Cloudbusting

There had, though, been an unsettling period when world governments almost opted for tax rebates on carbon emissions and increased energy-efficiency incentives for cars and homes. Thankfully, these schemes foundered at the last hurdle to allow for something bigger and better, something that would preserve the life you were accustomed to as well as showing her – the sun – that you had the measure of her ways. Meanwhile, carbon emissions kept on growing, the latest forecasts from the UN's International Panel on Climate Change estimating a maximum 6°C increase by the end of the century.

You also managed to help overrule some more mundane solutions. During 2009, governments and scientists had hoped that fluffy low-flying clouds could be utilized to reflect the sun's harmful rays away from the broiling land below. You spent months writing letters and releasing bogus online studies to discredit the efforts of John Latham of the National Centre for Atmospheric Research in Colorado, and Stephen Salter of Edinburgh University. They had designed a fleet of unmanned, self-propelled vessels that could cross the world's oceans and 'seed' clouds by firing a mist of seawater high into the air. A thousand of these ships could, they calculated, increase cloud cover by 4 per cent – enough to counter a doubling of carbon dioxide in the atmosphere. But the plan was

COLD COMFORT

riven with drawbacks. For a start, it was too cheap and relatively low-tech, which meant it would be quick to build and reliable to use. Worst of all was the gut-wrenching fear that it might even work. Your lobbying focused on the potential unpredictability of changing weather patterns that the scheme might induce, such as undesirable rainfall over drought-stricken African countries. Thankfully, negotiations descended into the usual squabbling between the developed and developing worlds and the idea was strangled at birth.

Sulphur the little children

Having proved beyond question that anything nature can do, humans can, quite frankly, do better, an idea has surfaced to create a man-made 'natural' catastrophe. Scientists noticed that when Mount Pinatubo in the Philippines erupted in 1991, the average temperature across the earth decreased by 0.6°C. The finger of blame pointed at the 10 million tonnes of sunlight-blocking sulphur that the volcano ejected into the stratosphere. You must establish formal contact with the eminent Professor Paul Crutzen, who won a Nobel prize in 1995 for his work on the ozone layer, and whose idea it was to replicate a Pinatubo-type explosion. His visionary plans involve hundreds of vehicles filled with sulphur – maybe aeroplanes or giant cannonballs, blasted into the stratosphere to create a cooling blanket that would block the sun's rays from reaching earth. You love it. You absolutely love it. You have meticulously researched the concerns of other scientists who claim that such a massive input of sulphur into the upper atmosphere could increase acid rain or damage the ozone layer. You learn that the Pinatubo explosion also caused a significant depletion in the ozone and, in the period following the eruption, the hole over the South Pole grew to a record size. You endeavour to track down Carnegie Institution climatologist Ken Caldeira and persuade him to keep these concerns to himself. Caldeira, a thorough and analytical scientist, can at times get bogged down with the most piffling of details. After all, he believes that putting sulphate particles into the stratosphere would actually 'destroy' the ozone layer. No, just damage it, my friend. A world of difference.

Crutzen's plan is perfect. True, global warming might slow, but ultimately the planet would fry to death. If all goes predictably wrong, it will take years for the particles to fall from orbit, years in which the planet may never recover.

Frankenstein foliage

Meanwhile, what you really must discourage are proposals that involve the mass planting of carbon-dioxide-absorbing trees. Scientists have proposed synthetic trees which, despite looking gratifyingly rubbish – they don't flower or leaf, and resemble 'goal posts with Venetian blinds' – can soak up carbon dioxide at dauntingly impressive levels. Klaus Lackner of Columbia University also needs to be silenced. His tree proposal at the annual meeting of the American Association for the Advancement of Science garnered valuable publicity for funding. Impressive calculations show that one of his trees could remove about 90,000 tonnes of carbon dioxide from the atmosphere in a single year – the output of more than 15,000 cars and a thousand-fold improvement on the poor old natural equivalent.

You write to every wildlife organization and newspaper, warning of a planned age of 'Frankenstein foliage' which, surely, will 'upset ecosystems by replacing healthy forests with floppy, disease-prone monocultures'. This, for a moment, actually sounds rather enticing, but sometimes even the defacing of aesthetics should take a backseat when it comes to truly screwing the planet.

On reflection ...

You also intend to fight the resurgence of plans to paint the ground white. Under these proposals, roads, oceans and deserts would be covered with reflective material, increasing the amount of sunlight reflected back into space and so cooling the planet. Far too sensible

COLD COMFORT

and achievable for starters and, worst of all, it might even prove value-for-money.

WHAT'S THE DAMAGE?

* Trials of painting surfaces white are abandoned after a series of crashes caused by drivers and pilots who are blinded by the light. **Maybe.**
* Attempts to fire giant mirrors into space suspended after series of technological hiccups. Costs soar to more than £5 trillion. **Likely.**
* Trials using artificial-cloud-making technology are declared a partial success. They do increase cloud, but also create an intense hurricane system that runs riot across the mid-west of America. **Plausible.**
* US and Chinese governments announce joint partnership to re-position the planet. By shifting its orbit away from the sun, they hope to cool it down. The energy of 5,000 million trillion hydrogen bombs is deemed sufficient to move earth's orbit by 1 mile and compensate for a doubling of carbon dioxide as a result. Nuclear warhead production goes into overdrive. **Unlikely.**
* Ideas come and go, promises are made but technological solutions fail. Emissions keep on rising. More promises are made. **Inevitable.**

Likelihood of any of the schemes having a major impact on climate change by 2015: 17%

Food 44 fright

The great supermarket swindle

AGENDA

* Put sham organic produce on shelves
* Corrupt consumer confidence
* Wreck rural reputations
* Import immoral alternatives

Not so long ago, your friends wouldn't have gone near the stuff. Wizened carrots, lumpy potatoes and gnarled courgettes that, frankly, look like they could be sold in an adult parlour. Now, everybody you know eats organic. They all feel pretty good about themselves. If you don't eat organic you don't care about the environment, right? In all honesty, it seems the notion is spiralling out of hand. Sales have whizzed past the £2 billion mark as supermarkets offer hundreds of such noble products. Locally grown, seasonally produced, fertilizer-free organic food is officially good for the planet.

FOOD FRIGHT

You have been increasingly worried for a while. Suddenly, the wonderfully destructive post-war farming practices that so unremittingly helped despoil the British countryside are under threat. Around 350 pesticides are currently allowed in conventional farming and an estimated 4.5 billion litres of chemicals are doused on British farms every year. But, one grey day soon, these figures will start to fall. Quite clearly, the organic revolution has to be stopped. And there is a way. What will happen when the whole thing is exposed as a con? What will happen when pesticide-soaked, habitat-devouring produce is revealed as being fraudulently passed off as organic? Yep, that will wipe the smug little smiles from their glowing, virtue-freckled faces.

Back to the soil

Frantically you assemble your fold-away foodie stall as the rain-clouds gather. It is your grand debut at a prestigious farmers' market in posh south-east London, and already the first sloane rangers are wandering over. You wipe the sleep from your eyes and force a thin smile. You are knackered. Most of the previous night was spent in your Bermondsey bed-sit opening sacks of supermarket own-label vegetables and smothering them in soil from the nearby municipal park. Some you singled out for special attention, chiselling holes in carrots' orange torsos, adding black ink dots or bending them into impossible curves. Now they look bad enough to be organic. Your potatoes look like lumps of coal. You've placed them in mouldy-looking sacks stolen from the back of a newsagent.

Attached to the side of the sacks are your home-made labels, depicting a rustic farmhouse with a quaint vegetable patch in the foreground. Behind are the outlines of some rolling hills and, beyond that, a vibrant rainbow. Arched above, in scarlet lettering, are the words: 'Butterscotch country holdings: Farm fresh: baby organic carrots, picked by hand from the highland loam. As nature intended.' The potatoes have been magically transformed into 'rich

Jersey spuds plucked from the renowned Channel soils'. All packaging is carefully marked: 'locally produced to strict organic standards'. The carrots cost you 44 pence per loose kilo. You will sell them for £1.30 per kilo. Your mass-produced potatoes have trebled in price from shelf to sack. By midday you have sold out of everything. Nice one! Thankfully, not one of your customers has asked to examine the certificate of authentication that legitimate organic producers are required to have to sell their produce. This organics lark is like taking candy from a baby.

Supermarket swizz

If you're really going to crush the organic movement, you must somehow shoehorn your fraudulent offerings into the aisles of the major supermarkets. Doing that would realize the deepest fear of the organic fanatics. Consumer confidence, notoriously fickle at the best of times, would be shattered. One major scare could signal the welcome end to the organics obsession. But, first, you will need a really dodgy supplier. You could track down one of the farmers who have falsely claimed to have organic accreditation. Contact trader Andrew Portch for advice. This grocer is a proven master of the fine art of organic fraudulence. The company he used to work for, Somerset Organics, has always stated on its website that their 'core philosophy is to produce and supply the highest quality certified organic food from the county of Somerset'. Portch himself was lying. He did not sell organics at all. But even he lacked the bottle to try and smuggle his stuff into the supermarkets.

To succeed in infiltrating the major supermarkets you will definitely need to improve your product. In fact, you need to actually grow your own vegetables, and no pesticides please – well, not just yet. Bear in mind that organics must be grown on a farm that limits chemicals and where crop rotation and compost are used. From now on, your labels must indicate the organic certification body with which you have registered. There are a few, but choose the Soil Association, which certifies the majority of organic produce. You cross your carrots and hope for the best. Inspectors pop down to inspect your beautifully kept garden and issue you with a compliance form. Everybody nods. You become a certified organic trader. Now you can print the Soil Association emblem on your carrots and a certification number. Remember that every organic producer is inspected at least once a year by the association. The Food Standards Agency also does spot checks, but at the time of writing prosecutions are rare.

Once certified, send a sample to the supermarket suppliers and hope for the best. In your favour is the fact that supermarkets will do anything to get their hands on organics. So much so that some allege there has been a lowering of standards and checks in the scramble to sate the demand of a sector growing at 30 per cent a year. Although you don't have enough produce to satisfy the demands of a major supermarket, you will argue that these are premium (priced) locally produced carrots that will reflect the supermarkets' willingness to help its nearby community. The supermarket, surprisingly, agrees to a trial after you have completed various forms tying your venture into an eye-wateringly merciless contract. Good job you're not in it for the money.

Your true colours

Now for the fun part. Immediately, you swap to using non-organic manure. At night you secretly spray the 'organic' carrots with delicious chemicals. Some of your potatoes come down with blight, so you use copper-based sprays which contain artificial chemicals. Then for the *pièce de résistance*. Using a syringe, you carefully inject your carrots with a squeeze of Allura Red AC (E129) and then Ponceau 4R (E124), an additive which has been linked to behavioural problems in children. Next, you wipe a couple of the vegetables with organophosphate pesticide (chlorfenvinphos), a substance which is banned but which was found in carrot samples as recently as six years ago. Manufacturers can use 5 per cent of certain non-organic food ingredients but still get away with labelling their products as organic, although your little extras might not be quite what they had in mind.

Now you have to be careful. The moment your produce arrives in the supermarket, call trading standards anonymously, offering them a tip-off that dodgy carrots, which you believe to contain chemicals, are being sold as organic. Also call the newspapers. Trading standards begin investigating. The supermarket denies the allegations, but three weeks of tests indicate that organic produce being sold by major multiples contains a cocktail of chemicals well above recognized safety limits. The supermarket launches its own internal inquiry. The Food Standards Agency instigates an independent investigation and the matter is debated in parliament. In a desperate move, the farming secretary seeks to reassure the public by stuffing an organic chicken burger into the mouth of his four-year-old daughter on live television. The ploy backfires when, just the next day, it is revealed that the poultry was flown 8,000 miles from Thailand by a Bangkok company that was recently fined after tests indicated the routine use of antibiotics on its chicken farms. The row escalates, and sales of organics dip sharply. All major supermarkets announce a routine check of organic suppliers in the country. Two months later, all are declared above board, but is too late. The damage is done. The dream has died. Cheaper, chemically soaked factory farming is back in vogue. Non-organic potatoes and carrots, which themselves can receive twenty different chemical sprays, are very much back on the menu.

In the meantime, you have two means of attack when it comes to destabilizing the organics movement. One is provided by the Soil Association itself, which has warned that much of our milk, cheese and yoghurt, as well as pork, comes from animals raised on genetically modified food. The association has raised fears that the public is being kept in the dark about farmers' and supermarkets' reliance on GM animal feed, simply because there is no legal requirement to mention it on food labelling. Secondly, if you must buy organic, then opt for the half of the sector which is flown in

FOOD FRIGHT

from across the world. People buy organics because they want the planet to be environmentally sustainable, yet they'll happily buy organics which have been imported 12,000 miles, creating a wonderfully large carbon footprint. Wipe out the local sector with imported organics: as you well know they taste quite delicious.

WHAT'S THE DAMAGE?

* Yet another study suggests that organic produce is no better for you than mass-produced alternatives. **Certain.**
* Cucumbers from Norfolk found with high concentration of banned chemical. **Imaginable.**
* Despite a major government crackdown on food fraud, just one prosecution is recorded during 2009, and that is for a farmer's grandmother who claimed to have got confused while selling apples at a Hertfordshire farmers' market. **Plausible.**
* Organics sector keeps on growing, breaking the £3 billion barrier by 2010. US chain Whole Foods opens Europe's largest organics hypermarket in Yorkshire. **Possible.**
* With British suppliers unable to meet demand, majority of organic products sold by supermarkets are imported from at least 2,000 miles away. **Likely.**

Likelihood of organics sector levelling off by 2015: 23%

Green gas 45

No-good do-gooder

AGENDA

* Pretend to be greener than thou
* The world hanging on your every word
* Sabotage the green movement

Single-handedly, you are going to hijack environmentalism. Yep, you're going to join the tofu-eating tree-huggers brigade. You will become a bitter, twisted sandal-ista whose doom-mongering and unrelenting preachery will marginalize the environmental movement. Your mission is to make the greenies irrelevant.

Already the writing is on the wall. Heed the words of the government's former chief scientist Sir David King, who described global warming as a greater risk than terrorism and warns that green activists are putting the fight against climate change at risk by wanting to take society back to the seventeenth century. There's also the founding father of the British environmental movement and chairman of the government's green watchdog, the

Sustainable Development Commission, Sir Jonathon Porritt, who has lambasted fellow environmentalists for being too 'narrow ... too depressing, too dowdy'. You must take it from here. Set up your own lobbying group to sabotage the sandal-wearers from within. You, my friend, are going to put the mental into environmentalism.

Factor in the feel-good

Already, your do-gooderism has received sponsorship from local firms. You are wacky, a bit out there, and you pretend you have what it takes to save the planet, so long as everyone works together. Your global organization operates from a one-man office in Watford. You have no business plan – being an environmentalist means you have no economic understanding and the financial acumen of a bedbug – but you do know you need a trustee. So, you write to James Lovelock, whose laboratory is in an old Cornish mill. This is the man who devised the Gaia hypothesis, a revolutionary theory that the earth is a self-regulating super-organism. The 88-year-old is considered a hero by the greens. More importantly, he unequivocally supports your hidden agenda, namely that lifestyle adjustments will do nothing to save the planet but might make some pathetic specimens feel better about themselves. 'Enjoy life while you can. Because if you're lucky it's going to be twenty years before it hits the fan,' he recently said.

While waiting for Lovelock to reply, you compile your first press release. It is titled 'Lighten up' and explains in needless detail that, by painting walls in a pale colour, homes require less artificial light, use less electricity and ergo save the world. Your second release explains how cleaning the back of your fridge can rescue us all from planetary apocalypse. 'Dusty coils increase energy consumption by 30 per cent.' Each release is stained with tears to symbolize the suffering that fat, rich Westerners are causing to the planet. Walk more. Eat by candlelight. Grow your own berries. Live in a long house. Stop breathing. Your stock-in-trade soon becomes the dissemination of tiny solutions to save the planet; nondescript yet ostentatious gestures that you, like your potential trustee, Lovelock, know will have absolutely no impact.

At the end of each press release is a box of statistics showing how the earth is falling apart at the seams. You are extra careful to promote only flawed figures documenting the collapsing ecology. Deforestation. Mass species extinction. Pollution. Over-use of chemicals. Over-consumption. Soil erosion. Global warming. All are indeed taking place, but your overblown figures are completely wrong. Already, you have garnered some media reaction for your 'Lilac lifeline: pale paint will save us all' exposé in the *Watford Observer*. You suspect that sales of lilac will dip sharply. Another check of press cuttings refers to your coil release with the desultory headline 'Earth in a strangling coil'. Its introductory paragraph reads: 'A Watford-based pressure group says that we will all die if we don't act now – starting with our fridges.' You write to ask for a correction, explaining that you are an *international* pressure group.

Gorilla tactics

With feedback distinctly underwhelming, you decide that sitting around in a flower-power shirt is getting you nowhere. Time to spice things up a bit. You hire five gorilla costumes and several inflatable bananas and, with four like-minded dowdies, stand outside the headquarters of a well-known multinational company that has interests in the Democratic Republic of Congo. Dressed as 'eco-chimpions', you chant and wave placards.

As security guards attempt to usher you from the steps, a scuffle erupts. You turn around to see a gorilla punching a middle-aged security guard on the chin. At that moment, a freelance photographer arrives, takes a picture of the scuffle and sells it to the *Sun*. Britain's most popular newspaper publishes it the next day with the headline: 'You couldn't ape it up.' Three weeks later, your group, minus the one on remand for GBH, climb on to the roof of parliament wearing matching hats made by an indigenous tribe at risk from tree-logging in Ecuador. You unfurl a banner

GREEN GAS

declaring that your (now planetary) liberation movement intends to f**k the system. Again, there is an unseemly grapple as police officers attempt to wrestle you before the gaze of the world's media. Eventually, you are arrested for trying to enforce existing government policy against the wishes of the government.

The fillip is that you are now official hero of the environmental movement, a *cause célèbre* among the eco-warriors. This time, your press release is pretty damn edgy and calls for the overthrow of capitalism through 'whatever means possible'. Its banality is enough to make the national press. Thom Yorke of Radiohead calls up to venture his support. You accept and smile approvingly, knowing that there is nothing better than a middle-class pop star telling the rest of us how to live. REM joins the bandwagon, saying it's the end of the world as we know it. Jeremy Clarkson refers to you in his *Sun* column as a 'mealy-mouthed environmental weird bear fool'. You have arrived, even if Lovelock has yet to reply.

During a subsequent guest-speaker appearance you reveal details of your next strategy to an audience of unreconstructed Sixties activists and upper-class Oxford undergraduates. You offer them £20 – a week's worth of lentils, tofu and organic wine straight from New South Wales – and they promise obediently to execute your plan. In early January 2009 London awakens to find that the statue of Winston Churchill has been defiled. Headgear from the Ecuadorian rainforest sits skewiff on his marble forehead. Graffiti sprayed on the base of the nearby Cenotaph reads: 'Eco-nazis, don't knock 'em'. Images of the man who rescued Britain from the black hole of true Nazism appear in every world newspaper. You are the new Banksy; the media has dubbed you the face of the entire green movement.

Fabricated figurehead
Overnight, you have been recast as the leader of a cult. Emails flock in from believers, messages of support from across the world. You take advantage and order all followers to wear flip-flops made from

African elephant hide. You preach constant self-flagellation, particularly for those caught carrying plastic bags, bottled water, or buying individually wrapped (gasp!) Quality Street chocolates. Your members are told they shouldn't date unless they are living together, thereby sharing lighting and heating. You even have temples – the wind turbines deep in the country from where you deliver sermons on dusty coils and the importance of lilac paint.

Mainstream groups like Friends of the Earth and Greenpeace issue statements to distance themselves from you, but no one listens. You *are* the green movement. You squarely blame big business and rich Westerners for all the planet's current troubles. And then you play your ace of spades. You publish your manifesto, which demands a slew of eco-taxes. You demand 20 pence per litre of petrol, tax on cheap flights, restrictions on 4x4s, a plastic-bag tax, fines if families flush their toilet more than once a day. 'The economic rape of the planet must be stopped if we are to let our children inherit a working earth,' you say, smiling smugly into the *Sky News* cameras.

Fallen idol

The paparazzi cannot get enough of you. They have already tracked down your mother, who says you were a lovely child who only ever seemed interested in killing worms and putting wingless flies under a magnifying glass when the weather was hot. But all good things come to an end. You call the tabloids to say that you will be outside McDonald's just after 3 p.m. At the allotted time you leave the restaurant clutching a Big Mac, a two-litre bottle of New Zealand spring water, and a genuine ivory carving. You wear a cotton shirt which came from the modern slave trade and was manufactured with an over-use of chemicals. The photographers snap you getting into an idling SUV you then drive aggressively towards Heathrow airport. You are heading to New York for a weekend designer-shopping spree. Inside the terminal, you make an impromptu press statement in which you explain that trying to

 **GREEN GAS**

save the planet is total bollocks, a massive waste of time and resources and that you didn't mean a word of it. 'Sorry, everyone.' You smile. You declare that it is far too late to cut greenhouse gases and that ethical shopping is a scam. And, finally, in the hall of Heathrow's new Terminal 5, besieged by the world's media, you borrow verbatim the words from the man for whose reply you are still waiting, 'Green is the colour of mould and corruption.' The whole concept is utterly jaded.

WHAT'S THE DAMAGE?

* The environmental movement becomes increasingly marginalized as people get bored of being told what to do. **Probable.**
* People realize that tiny sacrifices are making no difference whatsoever and environmentalism becomes seen as little more than faddism. **Likely.**
* An authoritative report says that the planet has passed its tipping point. In a pioneering statement it concludes that people might as well enjoy life while they can. **Predictable.**
* A series of riots erupt across Western capitals targeting affluent homeowners and businesses as a new activist movement decides to overthrow the state to try to save the earth. **Maybe.**
* Environmental issues move increasingly into mainstream politics. Government policies are amended accordingly but the world goes about its business like it always has. **Certain.**

Likelihood of environmental movement being seen as preachy by majority of public by 2015: 67%

Material 46 world

Dress to distress

....

AGENDA

* Launch your own clothes range
* Make chic really cheap
* Achieve must-have status
* As trends change, throw it all away

Fashion dictates. Like a sartorial Hitler, you either agree to its terms or are punished with public humiliation. No one can pinpoint the precise moment when clothes stopped being practical cover-ups from cold and shame and became the next weapon for environmental Armageddon. But it happened. Suddenly, you stopped donning fig leaves and began wearing clothes that accentuated your gender and identity. Sadly, clothing became the layer by which social or economic standing is graded. Fashion is so prolific, so fundamental, that society can hardly keep track of what is 'in' or 'out' but must follow nevertheless – blindly, in most cases.

 MATERIAL WORLD

Careers have been made but more have been broken by fashion. You are going to design clothing with the cachet of Cartier but at Primark prices. Your range will be ruinously heinous to ecology, but few will know and less will care. On launch day the queues will stretch three times around the block and Kate Moss, every kooky designer's wet dream and perhaps the most lusted-after fashion muse in the world, will be a huge fan of your fabulously chic line. In these heady days of over-consumption, making people buy what they don't need is the only game in town.

Cotton on

Kate Moss's Topshop range may have profited one of Britain's richest men, Sir Philip Green (who is worth nearly £5 billion), but last year it was revealed that the Asian workers who were manufacturing it were being paid less than £4 a day in Mauritanian factories. It certainly sounds like a money-spinner. Since Kate's look makes and breaks trends, replicating her style is an obvious must, but your range will differ in subtle but significant ways. It will be made from the finest cotton around. Cotton is the world's 'dirtiest' crop, using 16 per cent of all the world's insecticides, more than any other crop. And your batch will be sourced direct from the dirtiest fields of all, Uzbekistan, which, you are reliably informed, proudly boast some of the most toxicated cotton fields on the planet. Here, almost 1 kilogram of hazardous pesticides are applied for every few acres of cotton. The fibres for a single T-shirt demand an estimated 150 grams of pesticide to cultivate. In these damp fields the use of Aldicarb, a powerful nerve agent, and one of the most toxic pesticides ever created, is ubiquitous. Endosulfan – effortlessly capable of inducing coma, seizures, convulsions, and death among cotton farmers – will also be liberally used to further your fabulous range.

Naturally, your clothing must be cheap if you want it to sell. Over the last twenty years, the cost of high-street clothes has plunged, with many items costed at less than £20. To facilitate your low price tags, the cotton is flown from Uzbekistan to Delhi, destined for textile factories in the Shahpur Jat area of the city. Here child workers, some as young as ten, work in conditions as close to slavery as conceivably possible. They are paid as little as 50 pence a day for 14-hour shifts, to produce clothing for Western wardrobes. If their productivity wanes through fatigue or sickness, their gang masters can fine them or place the youngsters in manacles. Child's play.

Fashion chain

You're starting off your fashion empire with a modest range of four items: tea-dress, close-cut cigarette pants, a boyfriend blazer and a strictly limited edition over-sized clutch bag (made from unsourced leather from the back of an African crocodile). Your brand – called Shades of Green (SoG) – could be described as retro-style with a twist of modernity. The cut is flattering, the style classic, the colour varying shades of green to reflect your (dubious but unchallenged) eco-credentials. Think Moschino meets with Matalan. Savvy, sassy, chic and cheap. Bloody cheap.

One in every £4 spent on clothes in Britain goes on fashion, and if your label is going to affect the planet, then idiots must be able to afford it. A few buyers of course will know that cheap clothes come at a price, so be careful to obfuscate and sub-contract your supply line so that it is almost untraceable. Even your official book-keeping should be little more than a trail of smoke and mirrors. Undercover reporters will be alert for a whiff of slave labour so you must be careful to employ only the poorest of the poor, those who cannot risk their fragile livelihood in the name of exposure.

Of course, you'll require a celebrity model to wear the damn clothes. Moss's agent has not returned your calls so you settle on 22-year-old Agyness Deyn, British Fashion Awards model of the year and all-round purveyor of ethical wear. You call her agent, then her PR team, and send a sparkling brochure which features the phrase 'eco-chic' seventeen times and urges the need for a new age of enlightenment among shoppers. 'Look at the colour,' you say breathlessly on the phone, 'they are all green. Green. Geddit? These clothes were made specifically with the planet in mind,' you add. And you are not fibbing. Deyn, if she knows what's good for her, will undoubtedly say yep.

Brand bash

You must invite all the top fashion editors to a private viewing, where they will be offered mind-bending amounts of chemically reared pink champagne, very much non-organic salmon crudités and told to place orders for whatever they want from the imminent Shades of Green selection. It is a shameless ploy. Quite literally, you intend to buy their kind words. But the real *coup de grâce* is the launch party. Such bashes can crush or create a range before it goes public. A swinging, glittering affair crammed with the right faces is sufficient to ensure success. You have chosen The Hospital Club, a West End private-members club which is sufficiently elitist to conjure the illusion of self-importance and whose frequent guests include environmental ambassador Sienna Miller, who admits she cannot forgo flying but intends to take fewer baths.

Make sure to phone your paparazzi pals and run them through the guest list. Mention a stellar cast of A-listers, even if they're unlikely to turn up. Other essential guests are Katie Grand, editor-in-chief of the triannual *Pop* magazine; Alexandra Shulman, editor of *Vogue*; Rod Stanley, editor of *Dazed and Confused*, and its fashion editor, Katie Shillingford. Give high society a look-in by ensuring that *Tatler* editor Geordie Greig is chauffeured to the Hospital. Man-of-the-moment DJ Mark Ronson will be asked to lay some sounds. And don't overlook the goody bag. Sufficiently ostentatious, it can sell a party on its own. Aim to rival the gift bag dished out at the Oscars: scarlet lipsticks linked to the development of the blood disorder lupus, and £150 vials of parfum, man-made luxury fragrances proven to be toxic to the central nervous system. The crowning gift is blood-diamond jewellery, sourced from the

Côte d'Ivoire. This batch of diamonds helped to fund intractable conflicts and resulted in millions of deaths. Not surprising then that blood diamonds are banned. Investigations in 2007 found that many of the main UK jewellery retailers, including Cartier, Graff Diamonds, Fraser Hart, John Lewis and House of Fraser refused to provide any information on their diamond policy, with most companies having to be contacted several times.

Stylin'

With the first goody bags greedily received, you are already checking your phone in the hope that a high-street fashion chain or branded luxury wholesaler has contacted you with an opening bid to stock Shades of Green. Enter Primark. Three years ago it was rated the least ethical place to buy clothes and although it has since vastly improved, its grip on the bargain-hunting public exerts a powerful draw. Once Primark has made contact, it is time for that longstanding fashion-industry staple: the stunt, the final piece in the jigsaw. During the opening of her range, Moss modelled her clothes in the window of Topshop, and now Deyn will be asked to do the same in Primark's flagship store on Oxford Street. The difference will be the green body paint she will be sporting. With the launch timed for 6 a.m, unorderly queues begin to form the evening before. Contact the news desk of the *Evening Standard* and report scuffles between those eager to lay hands on Shades of Green. Pandemonium turns to panic as news leaks that there are just ninety-nine of each line available. Before most of the population arrives at work, a Shades of Green tea-dress flashes up on eBay for £700. The consumer clamour threatens to spiral out of control when the *Standard* reports a blazer literally ripped from the back of a young girl in Soho.

Amid ongoing hype, you launch the next edition of the range a month later, this time in a slightly different shade of olive-green. Deyn appears in full Shades of Green regalia on television shows *T4* and *Friday Night with Jonathan Ross*. Your range is now the most sought-after in Western Europe. By late 2009, SoG is the most must-have clothing collection on the planet. You decide to launch a second label and contact your friends in Delhi and those in Uzbekistan who may still be alive. Your public needs new, cheap offerings. All those who bought the first range have thrown the items away, and you hope they have ended up on a rubbish tip or landfill site. Yet in many ways there will be no end: your clothes will live for ever. The synthetic fabrics you wove into your poisoned cotton will not decompose, and this happy cocktail of chemicals will leach into the surrounding soil for many years to come.

WHAT'S THE DAMAGE?

* Mass-produced cheap clothes continue to dominate high-street fashion like never before. In a society where cost is all that matters, no questions are asked. **Certain.**
* Demand for luxury goods and ostentation grows as part of a large anti-environment counter-look. **Likely.**
* (Non-)ironical T-shirts saying 'People vs Global Warming' and 'Frankie says Fry' become must-have garments of the hot summer of 2010. **Probable.**
* Supermodels caught wearing slave-labour items from Uzbek fields. Media outrage, but the range sells out in record time. **Likely.**
* Less becomes more. Naturalism hits the high street and young men clad only in sardonic bowler hats and carrying empty briefcases becomes the cutting-edge look of July 2014. **Probably.**

Likelihood of eco-fashion dominating high street by 2015:
56%

47 Eco worriers

In poor 'ELF

....

AGENDA

* Terror tactics
* The backlash starts here
* Hound the Hummers

As an admirer of direct action, why not become a feared eco-terrorist and the leader of a cell linked to the Earth Liberation Front? You have probably been intrigued by the ELF ever since the early Nineties, when it was founded in Brighton during the British road protests. It soon spread to the rest of Europe and the US. Across the pond, the ELF became legendary as a group prepared to destroy property belonging to groups who have allegedly hurt the environment. This tactic has been so effective, in fact, that the FBI has classified 'ecotage' as the number-one domestic terrorism threat in the US and treats it on a par with al-Qaeda.

According to the FBI, the ELF (whose members are known as Elves) have been responsible for more than 1,200 criminal acts in the US, with the cost of their campaign close to £50 million. Creating an ELF cell would be one of the most effective means of tarnishing the mainstream green movement across the Western world. As leader, you'll be equipped to taint the environmental groups which campaign rather tiresomely against 'a profligate consumption of resources'. Official ELF policy is to 'create environmental sustainability' and at the same time to 'take all necessary precautions against harming any animal – human and nonhuman'. This seems unnecessary, so you won't be doing either.

ELF 'azard
They christened it the 'street of dreams', an ostentatious development in the quiet Washington state suburb of Woodinville, near Seattle. Its homes were billed as 'green', intended to tick every box of the well-heeled, ethically conscious families at which they were aimed. One night in March 2008, five large, half-built houses mysteriously caught fire. Neighbours described hearing explosives. In the sudden ferocious blaze, more than £3 million worth of damage was caused. A spray-painted bedsheet left at the scene read: 'Built green? Nope, black.' It was signed ELF.

Although the subsequent investigation could not find the perpetrators, the latest witch-hunt had begun. Commentators and libertarian bloggers used the attack as ammunition in their ideological war against green and left-wing campaigners. Environmentalism was back in the spotlight, dubbed misguided and irresponsible; an enemy of the state and of honest wage-earners everywhere. Some pointed out that burning down green homes would cause even more emissions, and accused the ELF of being hypocritical. It seemed your militant pals had done themselves more harm than good.

Run to earth
There's no such thing as bad publicity. You must capitalize on this situation and take ELF's campaign to greater, more damaging

heights than ever before. First, you need to make contact with the secretive, well-educated, and politically aware people behind this shadowy movement. It may be difficult. The official website of the ELF, which is clearly monitored by the authorities, does not respond to emails. In fact, you learn that, over the last six years, their website has only been updated sporadically, the once-frequent 'communiqués' to believers now conspicuous by their absence. There is no central ELF authority, no membership, no public meetings, and no mailing list. You cannot find an HQ or an official spokesperson. Like al-Qaeda, it is entirely decentralized.

Your attempts to track down Rod Coronado, who the FBI described as 'a national leader' of the ELF in the USA, reveal that he is in prison, having been charged in 2006 as part of the FBI's 'Operation Backfire', which ended in lengthy prison sentences for many suspected members. Instead, you manage to find some like-minded revolutionaries and self-fund a cell that will use the ELF as a banner for action. Your autonomous unit will be difficult for state and industry forces to infiltrate and destroy.

You turn to Dr Steven Best for advice. He co-edited a publication entitled *Igniting a Revolution: Voices in Defense of the Earth*, which includes essays by convicted environmental activists Jeff Luers and Noel Molland. Clean-shaven, tousle-haired Best also helped found the Animal Liberation's US press office, branded a terrorist organization by the FBI for its links with the ELF. You call the University of Texas, where Best is a professor of philosophy, and arrange a chat, taking precautions to ensure your phone is not tapped, because he is probably under surveillance. The risk is worth it: Best can offer you guidance and inspiration to shape your radical ecological ideal.

Suitably galvanized, you can get on with setting up your own group. A splinter group will be connected to an existing cell, known by the Charles Mansonesque title of The Family. The FBI believes it has shut down The Family after more than a dozen members based in Washington and Oregon were charged with causing seventeen fires, but you know better. The Family will be given the responsibility of liaising with ELF support networks in Belgium, Italy, North America, Poland, and Turkey.

Contacts are important, but be careful not to make yourself known to too many. An extraordinary state campaign is underway against green activists, some of whom now stand accused of arson offences from as far back as 2001. People in the States are paranoid; there is talk of a 'Green Scare', akin to the 1950s 'Red Scare' over communist infiltration. Your cell must consist only of trustworthy types; the threat of rogue people infiltrating your group cannot be overstated. Undercover agents may try and penetrate your group, a massive surveillance operation may attempt to catch you out, and beware of provocateurs hired by the state. Trust few and keep your head down.

ELF campaign

It is time for the campaign to begin. Nothing less than £1 million in damage will be considered a success. Initially, you sabotage power lines, which is a cinch. Next, you and your balaclava-clad believers burn down an SUV dealership, spraying 'Politics is pig meat' over the charred shells of vehicles. For a week you orchestrate a campaign which becomes known in the media as the 'Hummer bombings'. By the fourth night, security is stepped up at all SUV dealers in the Western world. Your terror campaign makes the front page of the *New York Times* and the *Guardian*. When a Manchester showroom featuring the Hummer H3 is engulfed in flames, images lead the BBC *News at Ten* bulletin. A fuzzy CCTV image of a wide-eyed figure in a mask appears on the internet. It is you. Newspaper articles refer to you as the sectarian SUV-slayer.

Your courage mounts every day. On the sixth night, two of your cell go on the rampage in the spiritual home of the urban SUV, southern California. They torch forty Hummer H2s. The phrase 'Hummer-burning' enters the American lexicon as a euphemism for extreme environmental behaviour. The Family claims responsibility. The next morning, you buy a Hummer to distract attention from you as a potential suspect.

Anything you can do ...

Any target that activists believe is detrimental to the environment is considered legitimate. And, as long as the attack is quick, organized and well-researched, you stand a good chance of getting away with it. One of the most celebrated ELF strikes happened in October 1998 in the ski resort of Vail, Colorado, which was scheduled to expand. The attack destroyed 885 acres of wilderness and caused up to £6 million worth of damage, the most expensive act of ecological sabotage so far. No suspects have yet been found.

You almost match this sum with an audacious attack on a chemical conglomerate on Teesside. With the world's media spotlight firmly upon your small band of anarchists, it's time for a masterstroke. You disseminate orders to torch a huge housing estate for the retired and rich on a Cornish Greenfield site. Your colleagues believe it is unoccupied. For years, the FBI have stated it is only luck that no one has been killed in fires started by Elves. That 'luck' is about to run out. The fires sweep through the complex, trapping a number of elderly on the upper floor. Seven die in the resultant blaze. The backlash is immediate. Overnight, sympathy for eco-activism – and environmentalism in general – evaporates, in the face of the monsters who put the sanctity of animal above human life. World leaders rally together, declaring that attacks against property and the elderly cannot be countenanced.

Fearing that the end is nigh, you close down your own treasured Family. But the threat to environmentalism will not so easily be contained. Commentators remark that your eco-terrorism has harmed the environment: the atmospheric effect of setting forty Hummers on fire released a harmful cocktail of chemicals, such as hydrogen cyanides, from the seat foam. One expert calculates that your attacks have released hundreds of tonnes of global-warming carbon dioxide – the equivalent of driving every one of those vehicles more than 10,000 miles. Hummers, whether they are being burnt or driven, are unfailingly good for your purposes. You go underground, leaving the mainstream environmental movement to defend itself. You are confident that, blood on its hands, it will struggle to recover its saintly image.

WHAT'S THE DAMAGE?

* Resurgence of ELF in Britain. First wave of attacks strike government departments amid claims that climate-change budgets have been cut. **Tenable.**
* Europe claims eco-terrorism is a more potent threat than al-Qaeda sympathizers. **Doubtful.**
* An audacious ELF attack in Oregon in autumn 2009 claims the movement's first victim after a Hummer explodes in a showroom and engulfs passing dog-walker in a fireball. **Probable.**
* FBI and Interpol officers round up more than forty ELF activists in a series of dawn raids after attack on high-profile Texas oil estate. **Possible.**
* Your good friend Dr Steven Best is banned from entering Britain for his latest book-signing tour. Weeks later he is arrested, after an email links him to British animal-protection activists campaigning against science laboratories in Oxfordshire. **Possible.**

Likelihood of ELF becoming a major European terrorist group by 2015: 26%

48

Brothers in arms

Oh! What a Lovely War

....

AGENDA

* Encourage insurgents
* Weather the (Desert) storm
* Play the great game

You love war. Most of all, you love how it moulds some of the most divine landscapes around. Forget broken bodies and human misery – you relish war for the way it never fails to f**k up the planet. During quiet hours, you often find yourself flicking idly through images of the WW1 Flanders battleground. Not a blade of grass anywhere, just miles of flat, mud-churned fields dotted with stumps of shattered trees like upright corpses.

Nothing can compete with war when it comes to reinforcing the natural order: who's going to care about species preservation when your very own is at stake? The poisons and toxic chemicals which accompany weapon production congregate in the atmosphere, ocean currents, and soil. Landmines are the perfect soldier: cheap, efficient, expendable, never hungry, never tired. They hang around harming the planet long after the fight has finished, and clearing them is laborious, dangerous, and thirty times more expensive than the weapon itself. Then there is the epitome of modern warfare, the nuclear bomb, proudly wreaking the most profound and persistent environmental damage during both its explosive release and its waste-rich aftermath. But the nuke is about to be toppled from its summit. A glorious new weapon is ascending the heights.

Push for Persian war

As he sat down for his nightly counsel with state television, President Ahmadinejad of Iran was in fine spirits, making bullish remarks about the Americans and insisting that Iran's controversial nuclear programme would continue, regardless of US threats. To the west of his country, Iraq's slide into internal conflict was gaining momentum. More than 4,000 US soldiers, almost 200 British, and at least 90,000 Iraqi civilians had died, victims of a mounting spiral of criminal, insurgent, and sectarian violence. But Ahmadinejad's concerns lay elsewhere. He knew that conflict in Iraq had fuelled rather than quelled support for Islamic extremism.

Ahmadinejad could never have dared imagine that Saddam Hussein would be overthrown and Shia co-religionists would come into power in Iraq. It was the first time that a Shia community had run an Arab country. Iraq, with its Shia majority, was, the president noted, a natural ally for the almost entirely Shia Iran. Now, Iran is ready to assume the role of spearhead state, leading the Muslim masses against international enemies who oppose their interests. You are delighted. You had been hoping that the environmentally debilitating war would continue in Iraq; now, another one looms. Checklist: convince the US to antagonize Ahmadinejad with threats of action. Tick. Impose several rounds of hardline sanctions against Iran. Tick. Leak detailed plans of military strikes against Tehran. Almost ticked. Let the Persian war begin.

HAARP on it

Nukes have had their day. With more than 150,000 troops stranded in Iraq, a full-scale military invasion seems too risky. What you need is a weapon which doesn't involve infantry, tanks or missiles yet can precipitate full-scale environmental Armageddon. Luckily, the US military, with a little help from your chums at British Aerospace Systems, have invented a wily system which, some analysts believe, could dwarf conventional- and strategic-weapon systems. The remote town of Gokona, Alaska, is home to the High-frequency Active Auroral Research Program (HAARP), an array of high-powered antennae that transmit, through high-frequency radio waves, massive amounts of energy into the upper layer of the atmosphere, the ionosphere. Official reports disagree, but some believe this system is capable of disrupting radio communications, interfering with electricity supplies, oil and gas pipelines, and even warping the mental well-being of entire regions. Some see HAARP as a potential weapon of mass destruction, capable of destabilizing agricultural and ecological systems, although it wasn't invented for this purpose. Others liken it to a gigantic heater which can burn long incisions in the protective ionosphere, allowing lethal radiation to strike the planet. It is thought that the system can trigger floods, hurricanes, droughts, and earthquakes.

You must persuade your twitchy but dependable comrades in the US defence department to unleash what you like to call 'weather modification'. As conflict with Tehran heightens, HAARP might be considered a non-conventional means of bringing those troublesome Iranians under control. What potential for generating precipitation, fog, and storms with which to whip Tehran into submission. Weather manipulation will ruin Ahmadinejad's agriculture and economy; his land will be at the mercy of food aid and imported staples from the US and your allies.

But mum's the word. So far, HAARP has been developed in secrecy, which is of course the *modus operandi* of BAE Systems. Such mystery has ensured that even discussion of the system is taboo. With the greenies obsessed with man-made greenhouse-gas emissions, you have identified the ultimate means to alter the weather. Meteorologists and military analysts remain mute on the subject. As for the Pentagon – well, quite rightly, it pours cold water on speculation that the enormous collection of transmitters, radars, and magnetometers in Alaska is some sort of super-weapon. Officially, HAARP is presented by the US Air Force as a 'research program'. They have no choice. After all, the 1977 international Convention ratified by the UN General Assembly bans 'hostile use of environmental modification techniques' characterized by the 'deliberate manipulation of natural processes'.

Fuel for thought

Man has yet to find a better use for oil than to fuel war. More than 300,000 barrels are used every day by the US in the war on terror, for vehicles and maintenance alone. Soldiers in Iraq use more than sixteen times the amount of fuel than did those in WW2. No wonder the Pentagon is the single largest user of oil in the world. In monetary terms, the Iraq war has cost the US up to $3 trillion. Even more, when you consider that every cent has served not only to divert funds away from measures to tackle climate change but also to widen the global rift with Muslim militants the world over. In short, to ensure that more wars, terrorist strikes and attacks against the planet are guaranteed.

More than five years of fighting in Iraq have left the country littered with chemical spills, hazardous material, and at least 311 sites which are poisoned by toxic, radioactive dust from depleted uranium shells. According to the UN, thousands of locations need assessing for environmental pollution. The mass-sabotage caused during the first Gulf War, when scores of Kuwaiti oilwells were set alight, hasn't materialized this time round. If someone were to repeat such an act it would desecrate the Mesopotamian marshes on the lower reaches of the Tigris and Euphrates rivers, Iraq's most prized environmental asset.

Hound the Afghan lands

There is another war which you are even more confident will keep on running at the earth's expense. This one is a cracker, one that will take huge efforts – and decades – to resolve; in other words, one that will continue until the world ends. Chat to British soldiers in Helmand province, southern Afghanistan, about the intractability of fighting, the miasma of tribal politics, terrorism, and the deaths of yet more friends. Already the conflict in Helmand has morphed way beyond the objective of crushing the Taliban. The Helmand valley has mutated into a geopolitical battleground for jihadists, a blooding ground for budding martyrs from across the globe. According to military predictions, it will be at least thirty years before war ends in Afghanistan and there is any hope of stabilization or reconstruction.

Visitors to Afghanistan remark on its natural beauty. Then they see the rusting hulks of Soviet tanks, the vast minefields that scar huge tracts of the country, and the villages crushed by US air raids. In many ways, the impact of modern warfare is best witnessed in the wild Afghan contours: decades of war have laid waste to its environment with such unerring efficiency that the UN believes reconstruction is compromised; more than half the forests in three Afghan provinces have been destroyed and, better still, almost no trees are detectable by satellite instruments in Badghis province, compared with a 55 per cent land cover three decades ago.

A team from the UN environmental programme's post-conflict assessment unit offers you a host of uplifting reasons to carry on fighting. It goes on, hailing Afghanistan's environment as being in a 'state of widespread and serious resource degradation: lowered

BROTHERS IN ARMS

water tables, dried-up wetlands, denuded forests, eroded land and depleted wildlife populations'. Even in the most remote areas, where there are snow leopards, Marco Polo sheep, wolves, brown bears, and Asian ibex, an upsurge in hunting has, with a certain panache, denuded populations.

War, you confidently predict, will continue both in Iraq and Afghanistan. Very soon, a third conflict will begin and, shortly after, the weather above Persia will begin behaving rather oddly. Forecasts will be bleak as the final, deadly assault against the planet gets well and truly underway.

WHAT'S THE DAMAGE?

* Leaked documents contradict US defence officials by revealing that HAARP was designed solely as a military weapon of mass destruction. **Maybe.**
* Iran increases frequency of Shia militia strikes against US troops. Pentagon threatens air attacks. Iran shrugs threats aside. **Foreseeable.**
* In late 2012 NATO generals admit Afghanistan may never be sufficiently stabilized. By then British casualties stand at four hundred, the same as those of Germany after it sent extra soldiers to Helmand in the summer of 2009. **Plausible.**
* Satellite photographs indicate that Tehran is, in fact, engaged in a nuclear-weapons programme. Other experts argue the images are doctored. US declares military action nonetheless. **Likely.**
* Britain withdraws its last troops from Iraq in 2012 amid heavy fighting in Basra as Iraqi forces struggle to pacify insurgency. **Possible.**

Likelihood of Iraq and Afghanistan remaining a conflict zone by 2015: 92%

Rock 49 squalid

Sing-along suicide

AGENDA

* Rock the planet
* Beat it
* The tour to end all tours

Imagine you are Simon Cowell. Actually, don't. Instead, conceive the idea that you are a consummate svengali with the creative nous to create the biggest, baddest band in the world. Ever. And when you say bad, you mean bloody wicked. Your rock outfit's legacy will be measured by its whoppingly impressive carbon footprint, its back catalogue fondly remembered for a not-so-ironic take on the planet's impending doom. The straightforward excesses will cast your band as the Mötley Crüe of the Thom Yorke years. The message is simple: to rekindle an age of hedonism, excess, and old-fashioned 'live for today' sensibilities, all wrapped up in an 'I don't give a f**k about starving polar bears' sentiment.

Fuller sh*te

At the moment you are missing an important ingredient. Four bright young things, aka the Band. But not for long. You must choose your band members carefully. Your quartet must have been raised by left-leaning, upper-middle-class parents and weaned on a formative diet of William Wordsworth and Charles Bukowski. To locate these types you could venture down to the BRIT School in Croydon, the state-funded birthplace of many successful acts including Adele, The Kooks and Katie Melua. Take in a talent show, search the school noticeboard for a spare four-piece indie rock band; do anything – just find a Band. Pick the skinniest, razor-cheek-boned, big-haired youths you can find and name them The Grassroots. Give them attitude, to spare. Book an East End venue, where some good-looking groupies with famous parents are bound to be wasting their unearned riches. You are not duly worried that The Grassroots are bottom of the bill. At this stage, it is less about the music and more about the people they snog. You have high hopes for lead singer, Ralph, and he doesn't let you down, succeeding in kissing Peaches Geldof, whose father had the foresight to criticize the recent Live Earth concert for lacking a 'final goal'. Ralph and Peaches spill lopsidedly on to a Soho street and get 'papped' by waiting photographers. The image appears in the tabloids with an unmerited mention of 'up-and-coming rock heroes The Grassroots'. Seizing the initiative, you book more dates.

After a particularly rowdy Barfly gig, your efforts to invite a senior A&R exec from 19 Entertainment, 'the world's most exciting and innovative entertainment company', finally pay off. This firm was founded by Simon Fuller, a man who knows a thing or two about marketing and is responsible for creating Brand Beckham and the Spice Girls. His guys love The Grassroots and their 'atitoood'. Fuller offers them a four-record deal. Their first single, 'I'm Greener ...' is produced by DJ Mark Ronson and is a huge success. The Grassroots are promptly signed up by clothing retailer Gap to model the following season's range.

Band goes big-time

'I'm Greener ...' becomes the soundtrack for British teen television drama *Skins* and The Grassroots hit the big time. The second release, 'Burning Up', garners yet more media attention by being released solely on USB stick, with Toast PR creating a suitably futuristic campaign to ensure that even the hippest kids are on board. By this stage, The Grassroots are playing larger venues, like Central London's Astoria and Koko in Camden. Now, somehow, Ralph bags Peaches' extremely beautiful younger sister, Pixie.

To offset increased sniping that they have sold out, The Grassroots play a round of secret gigs in underground dives such as the Notting Hill Arts Club and the drummer's bedsit in Whitechapel. Your lead singer is, once again, sick down his front, but manfully carries on performing. Ralph and Pixie begin hosting hilariously chaotic DJ sets together. They get very drunk in public. Days are rare that neither appear in the tabloid gossip columns. All of which is good – but you decide a change in gear is required. If The Grassroots are to have any lasting impact on the planet, then they are going to have to become considerably more famous.

You pay one of the less retiring members of the paparazzi to shove a telephoto lens in Ralph's face. As hoped, he thumps the photographer on the nose in front of the world. The next step towards total world domination is a slight cliché: your lead singer develops a hard-drug addiction. You escort him to the Priory after tipping off 'snappers' from the newspapers, just hours after he was spotted at Amy Winehouse's party in Camden. A slew of carefully placed stories appear, suggesting that 19 Entertainment are to sever ties with The Grassroots. Two weeks later, though, Ralph walks out of rehab, supposedly clean. Almost immediately, he gives a prearranged 'warts 'n' all' interview to *NME* in which the 21-year-old remembers to weep halfway through.

Recording begins for the difficult second album. As hype builds, a series of gigs are cancelled at the last minute to create a veneer of enigma and troubled genius. Finally, The Grassroots reappear.

ROCK SQUALID

But the line-up is different. They have become a supergroup. All original band members, bar Ralph, have been jettisoned due to 'mutual artistic differences'. Your efforts to piece together the biggest band in the world are about to come to fruition. On lead guitar, replacing Anthony from Muswell Hill, is Razorlight frontman Johnny Borrell, asked to join because he urged people to buy electric eco-scooters and then bought himself a 1000 cc Moto Guzzi bike (which he described as a 'monster revving beast'). Backing singers include Madonna and Sheryl Crow, Madonna because her carbon footprint touched an absolutely delectable 1,018 tonnes during her 2007 world tour; Crow because her hit 'Everyday is a Winding Road' was used to advertise carbon-munching Subaru 4x4s. The rhythm section stars Genesis. This three-piece squeezed their appearance at Live Earth, where other popstars joyfully highlighted climate change amid a wonderfully carbon-heavy forty-seven-stop world tour. You've always had the utmost respect for drummer and singer Phil Collins, especially since his trans-Atlantic Concorde dash during Live Aid 1985, so that he could play both London and New York on the same day. Bon Jovi agrees to play bass guitar. He flew from the UK to the US in his private jet to play in a New York stadium for the American leg of Live Earth. Finally, your backing band is the Red Hot Chilli Peppers, who, like all true rock stars, like to travel in style, producing a neat 220 tonnes of carbon dioxide with their private jet alone in just one half of 2007. You did not invite Thom Yorke.

Your band now has something for everyone, except for Damon Albarn of Blur, who accused the G8 international concerts of being 'too Anglo-Saxon'. The Super Grassroots' opening gig is at Wembley Stadium – it is voted the best gig of all time by svengalis the world over. Leaked information reveals that tour provisions clocked up 170,000 food miles – The *Ecologist* magazine describes it as an 'eye-popping feast that must surely be judged the most carbon-heavy food order for one group of people in the history of all humanity'.

Eat the world

The critics' response to new-look The Super Grassroots is phenomenal. It's the signal to arrange the Eat the World tour. Tickets, made from rare rainforest-sourced paper, are available only to those who make their own way. By 'make their own way', you mean private jet. Your egalitarian side finally wins, and you widen the criteria to allow tickets to those who can prove they have travelled at least 2,000 miles to attend. Cycle helmets are banned in case they are thrown into the crowd and cause head injuries.

The tour, naturally, is an ostentatious affair demanding enough electricity to power a small country for several months. In addition, you have requested the largest gathering of technicians, sound trucks, support staff, brass bands (with instruments sent separately by plane), and an eighty-strong troupe of dancers. The band plays to crowds of no less than ninety thousand in venues whose power supply does not come from renewable sources and whose facilities struggle to cope with the amount of waste and litter generated at each event. Each gig produces at least 1,000 tonnes of waste, which, as part of the venue contract, you have stipulated must not be recycled.

Each step of the 21-leg international tour (with one controversial performance in the Antarctic wilderness) eclipses the carbon emissions created by Live Earth, calculated by experts to be at least 31,500 tonnes. Your gigs produce the mileage total equivalent of an army. The total journey by fans and The Super Grassroots' huge entourage beats the 222,623 miles travelled by the superstars who performed at Live Earth. It goes without saying that you do not indulge yourself in carbon offsetting, even if Jon Bon Jovi so eloquently once said: 'We wrote a cheque, we took care of our footprint and raised awareness, blah blah blah.'

After the tour, you analyse Ralph's carbon footprint. Your lead singer has notched up a creditable 2,020 tonnes, 200 times the amount of carbon produced by the average Briton, more than 500 times that of the average African and, more impressively, twice as much as Madonna. During the final gig – in Sydney – Ralph mumbles incoherently about the importance of saving the planet in a speech that is later revealed to be stolen verbatim from Yorke. At least one commentator condemns him 'a massive, hypocritical fraud', a phrase that was previously used by a columnist in a description of Live Earth. Without fail, you start planning another tour. Only, this time, it really *will* eat the world.

WHAT'S THE DAMAGE?

* Live Earth II returns in the summer of 2010. Despite projected television audiences of two billion, less than a quarter tune in. **Likely.**
* Thom Yorke's carbon footprint is exposed as twice as large as the average Briton's after the singer is revealed to have two giant-sized jacuzzis permanently switched on in his attic. **Unlikely.**
* Eco-aware bands take charts by storm in 2011. A punk version of 'Greensleeves' by eco-anarchists is surprise hit of the summer. **Doubtful.**
* In response to carbon concerns, Madonna announces that she will no longer play foreign concerts. **Plausible.**
* Use of private jets among rich rock stars declines sharply in 2013. This is, however, nothing to do with eco-concerns; their popularity falls only after two crashes in the space of a month. **Possible.**

Likelihood of rock stars ceasing to use private jets by 2015: 15%

Blaze of 50 glory

It ain't over till it's over

AGENDA

* Choose to be cremated
* Plan your final fling
* Make a blazing exit

Face it: you are going to die. But don't despair, even on the way out there is still something you can do. In death, as in life, your proud promise to wreak carnage upon the planet can still be upheld. You can depart in a blaze of glory, your bloated body expunged in a final destructive burst of greenhouse gases. You will opt to be cremated in the hope that it is you who triggers the impending Armageddon, the burning of your carcass causing the pivotal release of carbon dioxide which finally pushes the planet past tipping point towards unstoppable meltdown. For you there will be no natural burial or touchy-feely return to the soil as part of the natural cycle of death, decay and rebirth. Get yourself down the local crem. It's better to burn out than fade away.

BLAZE OF GLORY

Burning up

You glance at your melting wristwatch and grin. It's one minute to midnight on the Doomsday clock. London has been engulfed by rising seawaters. Manhattan has sunk. Africa is wreathed in fire and famine. You decide to write your will. More than anything, you want your body incinerated in a final infernal halo of toxic global-warming gases. And so, in your neatest handwriting, you begin. First, you request that your teak coffin is made from the last remnants of surviving rainforest, scythed from an earmarked pristine plot on the Peruvian–Brazilian border and transported 7,000 miles to your final resting place. Next, you politely ask for the wood to be smothered in veneered chipboard and bonded with a formaldehyde resin, just like most cheap coffins. Formaldehyde is marvellous stuff, classified as one of the most hazardous compounds to the environment and an influential player in the happy hat-trick of acid rain, global warming and particulate air pollution. Around 1,000 grams of the poison is usually emitted per cremation, a figure you would happily die for. Naturally, your coffin should be smothered in the traditional coating of chemicals that never fails to unleash the toxins of hydrochloric acid, hydrofluoric acid and sulphur dioxide into the atmosphere. Government figures reveal that 12 per cent of dioxins coming into the atmosphere are from crematoria, which must surely be one of the most persuasive arguments for premature death that any authority has ever devised. Dioxins are described as among the most dangerous toxins around. So adept are they at toxifying that the World Health Organization once set their safe exposure at 10 pictograms per kilogram of body fat. As a pictogram is a millionth of a millionth of a gram, this, rest assured, is fine material.

As an extra request, you should ask to be embalmed. Considering you'll soon be dust, this is rather eccentric, but the practice prompts an additional generous release of formaldehyde and as such cannot really be ignored. Finally, you must make sure the undertaker knows that when you say you want to be burnt, you mean really *burnt*. Toast. You ask for the heat to be cranked up high. The usual temperature of a cremation is a cool 800°C, with a maximum of 1,150°C. You ask for 1,600°C. Minimum. This is important. The average cremation uses 285 kilowatt hours of gas and 15 kilowatt hours of electricity, around the same energy demanded by a living individual for an entire month. So be careful to state that you don't want your remains to arrive at the crem too early; it takes two and a half hours to heat a crematorium oven just to the 850°C point.

When you die you will be extremely fat. During your last few weeks on earth you will be so immobile that you might as well spend them inside your chemically coated coffin. But those extra body tissues sure are going to burn. This will be some bonfire. To achieve such an impressive body-mass index you must eat and drink so much gunk that something will surely give. First to go will probably be your molars. By the time you die you expect to have thirty-two teeth and twenty-eight fillings, all made from mercury, another brilliantly toxic heavy metal, poisonous to animal and plant life, and with an impressive knack of leaching into waterways and killing everything it comes into contact with.

Cremations release 16 per cent of the UK's mercury pollution and, unless tackled, are forecast to rise two-thirds by 2020. Your elephantine dimensions will play havoc with your joints towards the end and, ingeniously, you will have requested artificial joints made from plastic and therefore creating another valuable source of dioxins. You excitedly sign your will and smile. In the days before you started trying to f**k the planet, Britain recorded 650,000 deaths a year, 420,000 of which resulted in cremations. Now you are hoping for a full house, deliberately ignoring petty projections, which claim that by 2010 some 12 per cent of funerals will be classed as 'green'. For some time you have been telling anyone who will listen that woodland burials using ecological shrouds or eco-coffins provide nothing more than paupers' graves for the poor and misguided: 'If you want your mother dumped in a

BLAZE OF GLORY

cardboard box somewhere in a farmer's field, then fine, go ahead.' Gratifyingly, many in the funeral industry seem to be on message, and there are rumours that some have even attempted to block green burials, because they are cheaper and make undertakers less profit. At least one manufacturer of ecologically sensitive coffins came to your attention when he accused mainstream funeral directors of ignoring his low-cost coffins. In that instance, four thousand funeral directors were contacted about their knowledge of eco-friendly resting places, but just two responses were received. And anyway, with space such a priority on this crowded, denuded island, who wants to be crammed into a multi-burial site? People, understandably, get all squeamish when it comes to the thought of being piled on top of strangers in dank places.

Live for ever
You always dreamed of living for ever, a modern-day Peter Pan in an age of universal immortality, where the age of 500 marked the onset of adolescence. Currently, scientists are examining ways to put cellular ageing on hold. Tests on laboratory animals have produced reasonably positive results. Fruit flies, for instance, can double their natural lifespan and die healthy and vigorous. Famed Cambridge geneticist Aubrey de Grey believes that the first person to live to a thousand might be sixty already. Perhaps it won't be you, but there is news that gerontologists are fast at work on detailed plans to repair molecular and cellular damage to the body.

You know, of course, that even if humanity does ever achieve immortality, the planet hasn't got a hope in hell. Consider the 200,000 or so extra mouths who arrive on earth waiting to be fed each day, and it doesn't take de Grey to calculate that, with natural resources already being outstripped, the effect of this genetic breakthrough will prove compellingly destructive, leading to an overcrowded world where you would be competing with your own children in the workplace, where evolution itself would be neutered. Yes, man has constantly reinvented his environment, purely to find new ways to f**k the planet. Take the discovery of fire, invention of the wheel, electricity and nuclear energy – all of them, as you have proved, offer such unswervingly glorious potential for messing up the earth. Longevity, you hope, will be merely the latest. Perhaps, you ponder, the best plan of all might in fact be to just carry on as you are until the offer of immortality is unveiled. In the meantime, seize the day! Party on! You ain't never gonna die of boredom. And you might, just might, live for ever.

WHAT'S THE DAMAGE?

* Natural burials banned as space becomes a premium. Even triple-decker burial slots are considered a waste of a shrinking land. **Possible.**
* The Office of Fair Trading investigates complaints that eco-burials are being deliberately sidelined. **Maybe.**
* The age of immortality comes a step closer as scientists announce a genetic breakthrough in 2012. Tests on mice unlock the ageing process. Experts say four hundred is the new forty. **Remote.**
* Restrictions placed on cremation temperatures. Councils sanction only mass cremations, to cut down on greenhouse gases. **Likely.**
* Government eventually orders a lowering of crematoria temperatures to save energy. Bodies take five hours to burn. Grieving relatives complain at extended length of services. **Probable.**

Likelihood of cremations becoming increasingly popular by 2015: 8%

BLAZE OF GLORY

Resources

Belu Water
(The UK's first carbon neutral bottled water)
www.belu.org

British Beekeepers' Association
www.britishbee.org.uk

Campaign for Better Transport
www.bettertransport.org.uk

Campaign to Protect Rural England
www.cpre.org.uk

Carbon Footprint Ltd
www.carbonfootprint.com

The Carbon Trust
www.carbontrust.co.uk

Centre for Alternative Technology (CAT)
www.cat.org.uk

Department for Environment, Food and Rural Affairs (DEFRA)
www.defra.gov.uk

Earthwatch
www.earthwatch.org

Earth Day Network
www.earthday.net

The Ecologist Magazine
www.theecologist.org

The Eden Project
www.edenproject.com

Energy Saving Trust
www.est.org.uk

UK Environment Agency
www.environment-agency.gov.uk

Environmental Justice Foundation
www.ejfoundation.org

European Environment Agency
www.eea.europa.eu

The Fairtrade Foundation
www.fairtrade.org.uk

Forest Stewardship Council (FSC)
www.fsc.org
www.fsc-uk.org

Friends of the Earth
www.foe.co.uk

Greenpeace
www.greenpeace.org.uk

Green Helpline
http://green.energyhelpline.com

Green Party
www.greenparty.org.uk

National Biodiesel Board
www.biodeisel.org

Natural Resources Defence Council
www.nrdc.org

Peaceworkers UK
www.peaceworkers.org.uk

Rainforest Alliance
www.rainforestalliance.org

Recycle for London (Recycle-Now)
www.recycleforlondon.com

Royal Commission on Environmental Pollution
www.rcep.org.uk

Soil Association
www.soilassociation.org

Sustainable Development Commission
www.sd-commission.org.uk

Sustrans
(The UK's leading sustainable transport charity)
www.sustrans.org.uk

Tyndall Centre for Climate Change Research
www.tyndall.ac.uk

United Nations Educational, Scientific and Cultural Organization (UNESCO)
www.unesco.org

United Nations Environment Programme (UNEP)
www.unep.org

United Nations Framework Convention on Climate Change (UNFCCC)
http://unfccc.int

West Wales Eco-Centre
www.ecocentre.org.uk

The Wildlife Conservation Society
www.wcs.org

The Wildlife Trusts
www.wildlifetrusts.org

The World Conservation Monitoring Centre
www.wcmc.org.uk

World Health Organization (WHO)
www.who.int

The World Land Trust
www.worldlandtrust.org

World Wide Fund for Nature (WWF)
www.wwf.org.uk

Zoological Society of London
www.zsl.org

Acknowledgements

Mark: Thank you Morwenna for your love and patience and, of course, my wonderful family and friends.

David: I am grateful to my my lovely wife Katie, my best friend and my inspiration, and my gorgeous children Joe and Georgia who I hope will inhabit a better world. I remain forever indebted to my mother, Anita, now in her 81st year, who never stops believing in me and always reinforces that I can do anything if I put my mind to it. I would also like to thank my sisters Norma and Helen for always reminding me '...it ain't what you do it's the way that you do it ...'

... and of course two people without whom this book would have remained just another idea - David Dorrell for joining the dots and Mark Townsend for all his energy and painstaking research.

The authors would also like to thank:

Ivan Mulcahy - literary agent extraordinaire – for his good counsel, good humour and good company and publishers Jenny Heller and Ione Walder for believing in a new idea by two new authors.

The publishers would like to thank:

Sarah Day, Jeremy Tilston, Jody Barton, Richard Marston, Richard Mundon, Anne Rieley, Geraldine Beare and Trevor Kite.

Index

a

Abitibi-Consolidated **248**
Advisory Committee on Dangerous Pathogens **172**
Afghanistan **309-10**
African International Airways **97**
Afrimex **96**
Ahmadinejad, President **306**
Alaska Brokerage International Limited **70-1**
Alfa Group Consortium **178**, **180**
algae **140-5**
aluminium cans **189-90**
Amaggi Group **134**
Amazon rain forest **133-9**
American Association for the Advancement of Science **279**
Angel, Roger **276**
Animal Liberation Front (ALF) **22-4**
Animal Rights Militia (ARM) **23**
Arctic seed bank **51-6**
Asia Pulp & Paper (APP) **246-7**
Associated British Food and Beverages **236**
Attenborough, David **93**
aviation **215-20**

b

BAE Systems **197-8**
Ball, Dr Brenda **17-18**
Balls, Ed **225**
Baluyevsky, General Yuri **159-60**
Bartfield, Peter **71, 72**
Bate, Roger **260**
Bellamy, David **262**
Benedict XVI, Pope **242**, **272**
Benn, Hilary **168**
Best, Dr Steven **301**
biofuel **121-6**
birth rate **239-44**
Biwater **166-7**
book publishing **245-50**
bottled water **127-32**
Bottled Water Information Office (BWIO) **129-30**
bottom trawling **115-20**
Bovine Spongiform Encephalopathy (BSE) **138**
BP **179, 199**
Brazil **133-9**
BRIT School **312**
Britain **109-14**
British Airports Authority (BAA) **217-18**
British American Tobacco (BAT) **253, 255**

British Association of Dermatologists **154**
British Beekeepers' Association **17, 18, 323**
British Council **107**
British Nuclear Fuels (BNFL) **111**
British Soft Drinks Association **128**
Brown, Gordon **161, 218, 225, 231**
Bwindi Impenetrable National Park (Uganda) **95**

c

Caldeira, Ken **278-9**
Canadian Fisheries Department **72**
carbon offsetting **263-8**
Centre for Emergency Preparedness and Response (Colindale) **173**
Centre for Infections (Colindale) **173**
Centre for the Study of Carbon Dioxide **261**
Cernan, Eugene **204**
Cheney, Dick **180**
China **233-8**
China National Petroleum Corporation **42**
cigarettes **251-6**
CJD **138**
clams **37-8**
climate-change deniers **257-62**
coal **233-4, 235**

Coca Cola **128, 131**
Colony Collapse Disorder (bees) **16, 19**
Competitive Enterprise Institute (CEI) **260**
Congolese Democracy-Goma (RCD-Goma) **95, 96**
Conservation of Seals Act (1970) **73**
contraceptive pill **33-8**
Convention of Antarctic Marine Living Resources **49**
Coronado, Rod **301**
cotton **294-5**
Coulouthros family **178**
Country Guardian group **111**
cows **57-62**
cremation **317-21**
Crown Resources **178, 180**
Crutzen, Paul **124, 278, 279**

d

Danone **128**
Das Air **97**
Dashwood, Sir Edward **41**
deforestation **245-50**
Democratic Republic of Congo **94-8**
drug-trafficking **190-1**

e

Earth Liberation Front (ELF) **299-304**
eco-bulbs **152-6**
eco-towns and homes **146-51**

Eddington, Rod **218**
Edmonds, Noel **111**
E.J.Churchill **41**
Electricité de France (EDF) **224, 226**
elephants **39-44**
end of the world party **186-91**
endocrine disruptors **36-7**
Environ Energy Global **199**
Environment Agency **28, 34, 130, 155, 323**
environmental movement **287-92**
Environmental Research, Abersytwyth **60**
etinyloestradiol **35-6**
EU Biofuel Directive **122-3**
European Bank of Reconstruction and Development (EBRD) **179**
European Commission **117, 119, 194**
Export Credits Guarantee Department (ECGD) (Britain) **85**
ExxonMobil **261**

f

Family, The **301-2, 303**
FARC group **190-1**
fashion **293-8**
FBI **299-303**
feminisation of wildlife **33-8**
fertilizers **19**
fishing fleets **115-20**
FlyingMatters **217**

Food Standards Agency **284, 285**
Forest Stewardship Council **249, 324**
forests see Siberian forests
formaldehyde **318-19**
Friedman, Mikhail **178-9**
fungi **107**
  see also mycorrhizal fungus
FutureHeathrow **217**

g

gaz-guzzlers **192-6**
GC Rieber Skinn **71**
George C. Marshall Institute **260-1**
George F. Trumper Ltd. **43**
ghost netting **117**
Giesen, Stefan **178**
Global Diversity Crop Trust **53**
gorillas **93-9**
Great Apes Survival Project Partnership (Grasp) **97**
Green, Sir Philip **294**
greenhouse-gas emissions **58-62**
greenwash **197-202**
Grey, Aubrey de **320**
Griffin, George **172**

h

Hachette USA **247-8**
Halliburton **180**
Hamlyn **248**

Harmon, James **179-80**
Hawking, Stephen **207**
Health Protection Agency **173**
Health and Safety Executive **223**
Henman, Tim **148**
Henman, Tony and Janet **148-9**
High-frequency Active Auroral Research Program (HAARP) **307-8**
Hiroshima **157-8**
honeybees **15-20**
hormones **33-8**
House of Commons International Development committee **97**
Huhne, Chris **60**
Hutton, John **224**

i

Indonesia **122-6**
Ingham, Sir Bernard 111
Institute for Animal Health laboratories (Pirbright) **171-2**
International Bee Research Unit, Cardiff **20**
International Finance Corporation **137-8**
International Institute for Strategic Studies **162**
International Oil Spill Pollution Compensation (IOPC) Fund **180-1**
International Panel on Climate Change (IPCC) **141, 277**
International Policy Network (IPN) **258-9**
International Whaling Commission (IWC) **82**
Iraq **306-9**
ivory trade **39-40, 42-3**

j

Japan **82-5**
Japanese Knotweed **27-31**
Jones, Ian **143**

k

Karema (gorilla) **94, 95, 96**
Karmes Marine Fish Farm, Oban **22**
Kelly, Ruth **217**
Khan, Abdul Qadeer **161-2**
King, Sir David **259, 287**
KitKats **125**
krill **45-50**
Kyoto protocol **104, 258**

l

Lackner, Klaus **279**
Latham, John **277**
Life on Earth TV series **93**
Litvinenko, Alexander **159**
'Livestock's Long Shadow' (UN report) **59**
Livingstone, Ken **155**
Lovelock, James **288**
Luers, Jeff **301**

m

Maggi, Blairo **134-9**
Mare International **178**
Marine Bill (UK) **118**
Markov, Georgi **160**
Mato Grosso **134-9**
Merkel, Angela **124**
Meteorological Office **207**
Migraine Action Association **154**
mineral wealth conflicts **66-8**
Ministry of Defence (MoD) **76-9, 173**
Mitchell Beazley **248**
Molland, Noel **301**
Morris, Julian **258-9, 260**
Moss, Kate **294, 297**
motor cars **192-6**
Museveni, Yoweri **168-9**
Mutsambiwa, Maurice **42**
mycorrhizal fungus **87-92**

n

NASA **104, 131, 205, 206, 208**
National Centre for Atmospheric Research (Colorado) **277**
National Institute for Medical Research (NIMR) **173**
National Parks and Wildlife Management Authority (Zimbabwe) **41-2**
National Radiological Board **223**
Nestlé **125**
Nkunda, Laurent **95-8**
nuclear attack **221-6**
Nuclear Decommission Authority **225**
Nuclear Industry Association **225**
Nuclear Non-Proliferation Treaty **158-9**
nuclear plants **224-6**
nuclear war **157-63**
Nuclear Waste Trains Investigative Committee **223**
Nyakasanga Hunting Safaris **41**

o

obesity **209-14**
Ocean Nourishment Corporation (ONC) (Sydney) **142**
oestrogen **34-5**
oil **176-81**
O'Leary, Michael **58**
'Operation Backfire' **301**
organic produce **281-6**
Orion **248**

p

palm-oil **122-6**
Panasonic **236**
paper **245-50**
Para **135-8, 139**
peat swamps **122-3, 124, 125**
Philip Morris company **189, 253, 255**
Pires, Moacir **137**

planet-cooling **275-80**
Planktos **142-3**
plastic bags **227-32**
polar bears **63-8**
population growth **239-44**
Porritt, Sir Jonathon **288**
Portch, Andrew **283**
Prestige (ship) **177-81**
Primark **297**
Proctor & Gamble (P&G) **201**
Purves, Sir William **179**

r

rainforests **122-6**
religious teaching **269-74**
renewable energy **109-14**
Renewable Energy Foundation (REF) **111**
resources **323**
Rhrabacher, Dana **60**
Rio Tinto **189**
rock stars **311-16**
Rokke, Kjell Inge **47-50**
Rooker, Lord **18**
Roosmalen, Marc van **136**
Round Table on Sustainable Palm Oil **125**
Royal Astronomical Society **206**
Royal Commission on Environmental Pollution **48, 324**
Royal Horticultural Society (RHS) **53**
Royal Navy **79**
Royal Society for the Protection of Birds (RSPB) **112**

S

Sadat, Anwar **166**
safaris **40-1**
Sakhalin Energy **85**
salmon **21-6**
Salter, Stephen **277**
Schellnhuber, Sir John **207**
Science and Technology Laboratory (Porton Down) **173**
seals **69-74**
seed bank see Arctic seed bank
Sena, Edilberto **136**
SFK Pulp **248**
Shell **68, 85, 199**
Shell Transport and Trading Company **179**
Shenzhen Delux Arts Plastic **230**
Siberian forests **103-9**
Silverbacks see gorillas
Simon, Mary **65**
smoking **251-6**
Soil Association **283, 285, 324**
solar power **111**
Somerset Organics **283**
Sonar 2087 **77-9**
Southern Ocean **46-50**
soya **133-9**
space travel **203-8**
Stang, Dorothy **136**
Stern, Sir Nicholas **257**

Stevenson, William **117**
Suiping, Huaqiang Plastic **230**
supertankers **176-81**
Sustainable Development Commission **288, 324**
Svalbard Global Seed Vault **51-6**

t

Tesco **202, 231**
Thales Underwater Systems **79**
Toepfer, Dr Klaus **97**
transport industry **122-6, 192-6, 219-20**
trawling see beam trawling
Tyndall Centre for Climate Change Research (East Anglia) **207, 324**
Tyumen Oil (TNK) **179-80**

u

UK Centre for Medical Research and Innovation **173-4**
Unilever 125
United Nations (UN) **96-7, 142, 168, 180, 199, 228, 277, 308, 309**
Universe Maritime **178**
University of California **107**
urea **140-5**
US Marine Mammal Protection Act **66**

v

varroa mite **16-20**

viruses **170-5**

W

Walters, Sir Peter **179**
war **305-10**
water **164-9**
Watercourses Convention (1997) **168**
Werle, Hugo Jose Scheuer **137**
whales **75-80, 81-6**
Wicks, Malcolm **205**
Wiedeking, Wendelin **194, 195, 196**
wind and wave **109-14**
windmills **112-13**
World Bank **137-8, 235**
World Health Organization (WHO) **318, 325**

z

Zoological Society of London **77, 325**

First published in 2008 by Collins
an imprint of
HarperCollins Publishers
77-85 Fulham Palace Road
London W6 8JB

www.harpercollins.co.uk

Collins is a registered trademark of HarperCollins Publishers Ltd

12   11   10   09   08

5   4   3   2   1

Text © Mark Townsend and David Glick, 2008

The moral rights of the authors have been asserted.

All rights reserved. No part of this publication may be reproduced, stored in a retrieval system, or transmitted, in any form or by any means, electronic, mechanical, photocopying, recording or otherwise, without the prior written permission of the publishers.

Editorial Director: Jenny Heller
Project Editor: Ione Walder
Copy Editor: Sarah Day
Design Concept: Richard Marston
Design and Layout: Jeremy Tilston
Cover Design: Jody Barton

ISBN 978-0-00-727988-3

Printed and bound in Great Britain
by Clays Ltd, St Ives plc